A Guide to Distributed Digital Preservation

Copyright © 2010

This collection is covered by the following Creative Commons License:

Attribution-NonCommercial-NoDerivs 3.0 License

You are free to copy, distribute, and display this work under the following conditions:

(BY:)	**Attribution.** You must attribute the work in the manner specified by the author or licensor (but not in any way that suggests that they endorse you or your use of the work). Specifically, you must state that the work was originally published in *A Guide to Distributed Digital Preservation* (2010), Katherine Skinner and Matt Schultz, eds., and you must attribute the individual author(s).
($)	**Noncommercial.** You may not use this work for commercial purposes.
(=)	**No Derivative Works.** You may not alter, transform, or build upon this work.

For any reuse or distribution, you must make clear to others the license terms of this work.

Any of these conditions can be waived if you get permission from the copyright holder.

Nothing in this license impairs or restricts the author's moral rights.

The above is a summary of the full license, which is available at the following URL:

> http://creativecommons.org/licenses/by-nc-nd/3.0/legalcode

Publication and Cataloging Information:

ISBN:	978-0-9826653-0-5
Editors:	Katherine Skinner Matt Schultz
Copyeditor:	Susan Wells Parham
Publisher:	Educopia Institute Atlanta, GA 30309

TABLE OF CONTENTS

Acknowledgements vii
Road Map ix
Chapter 1: Preserving Our Collections, Preserving Our Missions 1
Chapter 2: DDP Architecture 11
Chapter 3: Technical Considerations for PLNs 27
Chapter 4: Organizational Considerations 37
Chapter 5: Content Selection, Preparation, and Management 49
Chapter 6: Content Ingest, Monitoring, and Recovery for PLNs 73
Chapter 7: Cache and Network Administration for PLNs 85
Chapter 8: Copyright Practice in the DDP: Practice makes Perfect 99
Appendix A: Private LOCKSS Networks 113
Glossary of Terms 127
Author Biographies 133

ACKNOWLEDGEMENTS

We wish to express our appreciation to the institutions and departments that provided various kinds of support to enable this publication. These include the Library of Congress through its National Digital Information Infrastructure and Preservation Program (NDIIPP), which has provided financial support for the founding and development of the MetaArchive Cooperative, and the National Historical Publications and Records Commission (NHPRC), which has provided funding for the Cooperative's transition from a successful project to a sustainable business. We are also grateful to Vicky Reich and the LOCKSS team who provided the technical framework for the Cooperative's network and who number among the most important pioneers of our generation, not just in digital preservation technology development, but in cultivating new ways for cultural memory organizations to collaborate to preserve our digital collections as community-based, community-owned, and community-led endeavors.

We would like to thank all of the contributors to this volume, first for their work in digital preservation, and second for their contributions to the substance of this book. This has been a group effort in every way and this volume depended upon the distributed knowledge and expertise of our community of contributors. Special thanks also goes to our copyeditor, Susan Wells Parham and the editorial expertise and input she has brought to this publication, and to Bill Donovan, whose fresh eyes relieved our own rather bleary ones as we finalized our work. We also greatly appreciate the efforts (and patience!) of our reviewer-editors—Vicky Reich and Tom Lipkis of the LOCKSS team, Richard Pearce-Moses of PeDALS, and Aaron Trehub of ADPNet and MetaArchive.

Finally, we would like to extend a special thanks to the distributed community of readers who have assisted us along the way and to our future readership, from whom we hope to hear in later editions of this book. The distributed digital preservation field is quickly expanding and as we state repeatedly herein, other technical solutions and organizational approaches are emerging and maturing rapidly. We look forward to working with you all in the growing community of cultural stewardship that we will collectively foster.

Katherine Skinner and Matt Schultz
February 2010

ROAD MAP

A Guide to Distributed Digital Preservation is intentionally structured such that every chapter can stand on its own or be paired with other segments of the book at will, allowing readers to pick their own pathway through the guide as best suits their needs. This approach has necessitated that the authors and editors include some level of repetition of basic principles across chapters, and has also made the Glossary (included at the back of this guide) an essential reference resource for all readers.

This guide is written with a broad audience in mind that includes librarians, curators, archivists, scholars, technologists, lawyers, and administrators. Any resourceful reader should be able to use this guide to gain both a philosophical and practical understanding of the emerging field of distributed digital preservation (DDP), including how to establish or join a Private LOCKSS Network (PLN).

Please note that our title is "A Guide," rather than "The Guide," to distributed digital preservation. We are practitioners, and as such, are covering in this book the territory that we know well. While the LOCKSS software is certainly not the only approach to DDP, it has reached a high level of maturity in the library, archives, and museum communities. We have chosen here to focus some of our attention on this technological approach because we use it ourselves and because we believe that it is readily available to groups of institutions that wish to quickly begin preserving their collections in geographically distributed preservation networks. We are excited to see other promising technologies reach similar levels of maturity, and we hope that similar documentation of these methodologies and approaches will be offered to the extended community in the future, either through other books and articles or through future editions of this guide.

The Chapters

Chapter 1: Preserving Our Collections, Preserving Our Missions

Martin Halbert and Katherine Skinner provide a philosophical base for cultural memory organizations' need to participate in distributed digital preservation solutions as community-owned and community-led initiatives. This chapter will be useful for all readers, particularly those with questions about the value of collaborative engagement in the digital arena for cultural memory organizations.

Chapter 2: DDP Architecture

Katherine Skinner and Monika Mevenkamp explore the architecture of DDP networks, focusing primarily on the Private LOCKSS Network (PLN) and its central elements. This chapter provides a foundation for all of the chapters that follow and as such, is highly recommended for all readers.

Chapter 3: Technical Considerations for PLNs

Beth Nicol focuses on some of the core technical decisions that different PLNs have made based on the needs of their members. This chapter will be most useful to administrators and technologists who are thinking about producing or joining a DDP, especially one that uses the LOCKSS software.

Chapter 4: Organizational Considerations

Tyler O. Walters provides an overview of the administrative and organizational apparatuses that are available to DDP networks and uses a series of case studies to explore the operational decisions made by existing PLNs. Administrators who are considering hosting or joining a DDP network of any type will find this chapter a helpful guide.

Chapter 5: Content Selection, Preparation, and Management

Gail McMillan and Rachel Howard offer a set of best practices for policies regarding the selection, preparation, and management of digital content for preservation purposes. Though they draw upon PLNs in their featured examples, the chapter may be applied more broadly to many DDP initiatives. As such, it will be of great use to librarians, curators, and archivists, as well as administrators, who are seeking to ready their collections for preservation.

Chapter 6: Content Ingest, Monitoring, and Recovery for PLNs

Katherine Skinner, Matt Schultz, and Monika Mevenkamp describe the central elements of the PLN-based strategy: content ingest, monitoring, and recovery. This detailed chapter will be most useful to librarians, archivists, curators, and technologists who are participating as content contributors and as a preservation site for a PLN.

Chapter 7: Cache and Network Administration for PLNs

Matt Schultz and Bill Robbins provide an overview of the different network responsibilities incurred through two different approaches to running a PLN: those that depend upon the LOCKSS team to provide central administration services and those that form independently of the LOCKSS team and self-administer their networks. They also describe the basic functions undertaken by local systems administrators in each approach as they run a cache for the network. This chapter will be most useful for administrators and technologists of institutions that seek to host a network or maintain a cache for a network.

Chapter 8: Copyright Practice in the DDP: Practice makes Perfect (or at least workable)

Dwayne K. Buttler examines the legal implications of copying and distributing digital content for preservation purposes, focusing on the U.S. context. He looks at the origins of copyright law, then delves into the specifics of the DDP methodology based on its three main components: contributing content, preserving content, and restoring content. For each of these stages of preservation work, he provides a list of useful questions that content contributors can refer to when determining their rights to use DDP methods to preserve their collections.

A Guide to Distributed Digital Preservation

Edited by Katherine Skinner and Matt Schultz

Educopia Institute
Atlanta, Georgia

Chapter 1: Preserving Our Collections, Preserving Our Missions

Martin Halbert (University of North Texas)

Katherine Skinner (Educopia Institute)

As the collections of cultural memory organizations become increasingly digital, preservation practices for these collections must likewise turn to digital techniques and technologies. Over the last decade, we have witnessed major losses of digital collections, both due to large-scale disasters (e.g., hurricanes Katrina and Rita, the 2003 power grid failure of the northeastern U.S. and southeastern Canada) and more isolated, local-level events (media failures, human errors, hacker activities, and smaller-scale floods and fires). From these losses, we are learning how vulnerable our digital collections are and how urgently we need sustainable digital preservation practices for our cultural stewardship community.

Paradoxically, there is simultaneously far greater potential *risk* and far greater potential *security* for digital collections as compared to physical and print collections. Risk, because such collections are as ephemeral as the electrons with which they are written, and can be catastrophically lost because of both technical and human curatorial failures much more easily and quickly than our physical and print-based holdings. Security, because digital collections, unlike physical artifacts, can be indefinitely reproduced and preserved with perfect integrity and fidelity. For all intents and purposes, anything less than perfect continuity of digital collections implies complete corruption and loss of data. Thus we as cultural stewards must create fail-safe methods for protecting and preserving those collections that we deem to be of sufficient cultural and historical importance.

The apparatuses, policies, and procedures for preserving digital information are still emerging and the digital preservation field is still in the early stages of its formation. Cultural memory organizations are experimenting with a variety of approaches to both the technical and organizational frameworks that will enable us to succeed in offering the perfect continuity of digital data that we seek. However, most cultural memory organizations are today underprepared for the technical challenges incurred as they acquire, create, and preserve digital collections.

As a result, troubling trends are already developing within our community that may be counterproductive to our overall aims. For example, many cultural memory organizations are today seeking third parties to take on the responsibility for acquiring and managing their digital collections through contractual transfer and outsourcing of operational arrangements. The same institutions would never consider outsourcing management and custodianship of their print and artifact collections; the very notion is antithetical to cultural memory organizations, which exist specifically for the purpose of preserving and maintaining access to these collections. Yet institutions are today willingly giving up their curatorial responsibilities for their digital collections to third parties, precisely at the time that these digital collections are becoming their most important assets.

The central assertion of the MetaArchive Cooperative, a recently established and growing inter-institutional alliance, is that cultural memory organizations can and should take responsibility for managing their digital collections, and that such institutions can realize many advantages in collaborative long term preservation and access strategies. This assertion is based both on the shared convictions of our members and on the successful results that MetaArchive has achieved in recent years through coordinated activities as a cooperative association.

Authored by members of the MetaArchive Cooperative, *A Guide to Distributed Digital Preservation* is intended to be the first of a series of volumes describing successful collaborative strategies and articulating specific new models that may help cultural memory organizations to work together for their mutual benefit.

This volume is devoted to the broad topic of distributed digital preservation, a still-emerging field of practice within the cultural memory arena. Digital replication and distribution hold out the promise of indefinite preservation of materials without degradation, but establishing effective processes (both technical and organizational) to enable this form of digital preservation is daunting. Institutions need practical examples of how this task can be accomplished in manageable, low-cost ways.

We have come to believe that the use of the LOCKSS (Lots of Copies Keep Stuff Safe) software developed by the Stanford University Libraries-based LOCKSS team for collaborative digital preservation purposes is one effective, practical, and affordable strategy that many collaborative digital preservation initiatives may wish to consider. Portions of this volume thus delve into the specific topic of deploying the LOCKSS software to create Private LOCKSS Networks (PLNs). These are membership-based geographically distributed networks that are dedicated to the long-term survival of digital archive. The MetaArchive Cooperative has successfully operated a shared digital preservation infrastructure based on this model for more than six years, and has advised other groups in the implementation of similar networks. With eight known international examples in operation today, PLNs are arguably the first well-established approach to distributed digital preservation within the cultural memory arena. This is not to say that PLNs are the only approach to distributed digital preservation. Indeed, we hope that this book will someday stand as one among many guides and that practitioners creating other promising frameworks will produce documentation about the technical and organizational approaches they use in order to foster additional communities of development in this important field.

The remainder of this introductory essay will serve to outline the early consensus on the emerging field of distributed digital preservation (DDP) and the rationale for networks based on the use of the LOCKSS software that has been modeled by the MetaArchive Cooperative, the Alabama Digital Preservation Network, and other PLNs over the last six years. Our hope in this guide is to begin to advance the current conversation on these topics among cultural memory organizations, topics that we believe are central to the continued vitality and success of such institutions.

AT-RISK CONTENT AND THE EMERGING DDP FIELD

Most cultural memory organizations do not yet have a digital preservation program, although most are aware of their need for one. According to the 2005 Northeast Document Conservation Center (NEDCC) Survey by Liz Bishoff and Tom Clareson, 66% of all cultural memory institutions report that no one is responsible for digital preservation activities, and 30% of all archives have been backed up one time or not at all.[1] These statistics should be staggering and alarming for leaders of cultural memory organizations that expend major funding on the creation and acquisition of digital collections. As the service programs in cultural memory organizations become increasingly focused and dependent on long-term access to these digital collections, the gap that exists in preservation efforts for these collections becomes all the more critical to address. Yet, the previously mentioned gap in our collective understanding of how to respond to this challenge is a reality; many institutions simply do not know what to do.

What is at stake if we do not build this proposed new field of digital preservation? The National Digital Information Infrastructure Preservation Program (NDIIPP) has highlighted the scope of the collections now at risk:

> Technology has so altered our world that most of what we now create begins life in a digital format. The artifacts that tell the stories of our lives no longer reside in a trunk in the attic, but on personal computers or Web sites, in e-mails or on digital photo and film cards.... When we consider the ways in which the American story has been conveyed to the nation, we think of items such as the Declaration of Independence, Depression-era photographs, television transmission of the lunar landing and audio of Martin Luther King's "I Have a Dream" speech. Each of these are physically preserved and maintained according to the properties of the physical media on which they were created. Yet, how will we preserve [the following] essential pieces of our heritage?
>
> - Web sites as they existed in the days following Sept. 11, 2001, or Hurricane Katrina?
> - Web sites developed during the national elections?
> - Executive correspondence generated via e-mail?
> - Web sites dedicated to political, social and economic analyses?
> - Data generated via geographical information systems, rather than physical maps?

- Digitally recorded music or video recordings?
- Web sites that feature personal information such as videos or photographs?
- Social networking sites?

Should these be at a greater risk of loss, simply because they are not tangible?[2]

A great deal of content is in fact routinely lost by cultural memory organizations as they struggle with the enormous spectrum of issues required to preserve digital collections, including format migration, interoperability of systems, metadata to make the collections intelligible, and a host of other challenges. If this range of challenges was not enough, best practices for the most basic requirement of all are still poorly understood, namely how to ensure the long-term continuity of the bytes of data that fundamentally comprise digital collections.

Backups versus Digital Preservation

There are some that would dispute the above statements. Backing up the content on servers to tape and other static media is a long-standardized component of system administration. Why do we differentiate data backups from digital preservation programs? As the Joint Information Systems Committee (JISC) wrote in 2006:

> Disaster recovery strategies and backup systems are not sufficient to ensure survival and access to authentic digital resources over time. A backup is short-term data recovery solution following loss or corruption and is fundamentally different to an electronic preservation archive.[3]

Backups have always been tactical measures. Tape backups are typically stored in a single location (often nearby or collocated with the servers backed up) and are performed only periodically. As a strategy, backups are designed to address short-term data loss via minimal investment of money and staff time resources. While they are certainly better than nothing, backups are not a comprehensive solution to the problem of preserving information over time.

Digital preservation is strategic. A digital preservation program entails forming a geographically dispersed set of secure caches of critical information. A true digital preservation program will require multi-institutional collaboration and at least some ongoing investment to realistically address the issues involved in preserving information over time. It also requires the creation and maintenance of preservation policies and procedures that guide the long-term curation of digital collections.

WHY DISTRIBUTED DIGITAL PRESERVATION?

In the context of this critical need for a new consensus on how to preserve digital collections, a growing number of cultural memory organizations (including those of the MetaArchive Cooperative) have now come to believe that the most effective digital preservation efforts in practice succeed through some strategy for distributing copies of content in secure, distributed locations over time. This conceptual strategy is a straightforward carry-over of the practices that in the chirographic (handwritten) world of antiquity enabled scholars to preserve content through millennia of scribal culture. But in the digital age this strategy requires not only the collaboration of like-minded individuals, but also an investment in a distributed array of servers capable of storing digital collections in a pre-coordinated methodology.

A single cultural memory organization is unlikely to have the capability to operate several geographically dispersed and securely maintained servers. Collaboration between institutions is essential, and this collaboration requires both organizational and technical investments. Not only a pre-coordinated technological solution, but also strong, long-term inter-institutional agreements must be put in place, or there will be insufficient commitment to act in concert over time. The following quote from a joint National Science Foundation (NSF)/Joint Information Systems Committee (JISC) study captures the opportunity presented by this situation succinctly:

> The increased number and diversity of those concerned with digital preservation—coupled with the current general scarcity of resources for preservation infrastructure—suggests that new collaborative relationships that cross institutional and sector boundaries could provide important and promising ways to deal with the data preservation challenge. These collaborations could potentially help spread the burden of preservation, create

economies of scale needed to support it, and mitigate the risks of data loss.[4]

The experience of the MetaArchive Cooperative echoes this sentiment, namely that any effective implementation of distributed digital preservation requires both a robust technical infrastructure and strong inter-organizational arrangements. By "robust technical infrastructures" we especially mean strategies combining geographic distribution to multiple locations and security of individual caches, a combination of approaches that maximizes survivability of content in both individual and collective terms. Maximizing security measures implemented on individual caches reduces the likelihood that any single cache will be compromised. Distribution reduces the likelihood that the loss of any single cache will lead to a loss of the preserved content. This combination of strategies enabled documents to survive over millennia in the scribal world. We do not yet know if they will have similar results in the digital world, but they offer the most promising strategy to date.

A CALL TO ACTION

A central purpose for the MetaArchive Cooperative as we wrote this guide was to build documentation to help strengthen the distributed digital preservation and Private LOCKSS Network (PLN) communities and to encourage institutions to create and engage in collaborative preservation strategies with each other. Cultural memory organizations understand preservation issues in ways that other entities do not, and as a community, we must value both the training and the mindsets that librarians and curators bring to the virtual table as we pioneer solutions for preserving our digital collections. Philosophically and practically, our mission as museums, libraries, and archives is twofold: to provide access to and to preserve those objects deemed by curatorial experts to be most important for current and future generations. We need to provide preservation services more urgently than ever before due to the vulnerability of our digital assets. If cultural memory organizations do not take an active role in the preservation of our own collections, and rather cede this responsibility to external agents to do it for us, we run the risk of undermining our own stability as institutions.

To put that more plainly, to outsource one of our two key missions in the digital medium is to begin acting as brokers rather than curators—a dangerous step in any time, but particularly so in one so fraught with economic tensions. If we want to continue serving as cultural stewards of the digital age, we must be active players, not passive clients of third-party preservation services. We learned a difficult lesson in the university library community in the 1990s when we chose to pay for temporary access to major journals that were digitized by corporate entities rather than creating and owning that digital content for ourselves; we cannot afford to repeat that mistake in the realm of preservation.

One of the greatest risks we run in not preserving our own digital assets for ourselves is that we simultaneously cease to preserve our own viability as institutions. One of the costs to us as institutions if we ignore, postpone, or outsource our duty as cultural stewards of a digital age is that other institutions will fill the gap that we leave, likely to the detriment of our institutional community and with it, to our cultural memory.

So how can we affordably approach digital preservation for ourselves? This guide provides a sketch of the current DDP landscape and the promising models that are emerging within it, and outlines for interested institutions a pathway forward.

COLLABORATIVE APPROACHES TO PRESERVATION

Organizationally, the DDP value proposition is both simple and radically different from that of most businesses. It advocates seeking to reduce both short- and long-term costs by investing in commonly owned solutions. DDP networks require deep and long-term commitments from their members in order to serve the preservation needs for which they are designed. We need new technical and organizational models for undertaking this collaborative work as we work to sustain our activities.

In the following chapters, readers will learn about different approaches to DDP. One of these approaches has been pioneered by groups of institutions using a common technical framework, that of the Private LOCKSS Network (PLN). However, the resources required to run these PLNs varies widely depending upon the specific decisions made by each PLN at both a technical and organizational level. Some PLNs are run within existing organizational structures and administered by the Stanford LOCKSS team. Other PLNs are independent entities. Some PLNs use the standard tools provided by the LOCKSS team; others couple these tools with additional preservation tools and services to meet more specific preservation needs. PLNs, then, can be created with different levels of technical and administrative complexity. Each approach has its pluses and minuses, but all have been shown to accomplish the central aim of distributed digital preservation. It is up to the institutions that host and participate in a PLN to decide which approach will work best for them.

We hope that this guide will help to disseminate information about some of these emerging models and in doing so, assist other cultural memory groups in their efforts to create distributed digital preservation networks.

ENDNOTES

1. See http://www.ils.unc.edu/digccurr2007/slides/bishoff_slides_8-3.pdf (last accessed 01-29-2010).
2. See http://www.digitalpreservation.gov/importance/ (last accessed 07-21-2008).
3. JISC. "Digital Preservation: Continued access to authentic digital assets." (November, 2006)
4. Berman and Schottlaender, "The Need for Formalized Trust in Digital Repository Collaborative Infrastructure, NSF/JISC Repositories Workshop (April 16, 2007) http://www.sis.pitt.edu/~repwkshop/papers/berman_schottlaender.html (last accessed 01-20-2010).

Chapter 2: DDP Architecture

Katherine Skinner (Educopia Institute)

Monika Mevenkamp (Educopia Institute)

OVERVIEW

This chapter provides an overview of the basic technical architecture of a distributed digital preservation (DDP) network, distilling that framework down to its central components. It then focuses on one popular open source technical infrastructure that supports distributed digital preservation practices: the Private LOCKSS Network (PLN). It presents the fundamental elements of the PLN architecture and details how they interact to form a preservation network. This chapter provides the foundation for more in-depth discussions of the PLN framework that are featured later in this book, including in Chapter 3: Technical Considerations for PLNs, Chapter 6: Content Ingest, Monitoring, and Recovery for PLNs, and Chapter 7: Cache and Network Administration for PLNs.

THE BASICS: DIGITAL PRESERVATION AND OAIS

This guide is not intended as a starting place for those who are new to digital preservation, and thus will not provide basic-level information about digital preservation. We do wish to begin by pointing individuals who lack this basic familiarity to the ISO standard that provides grounding to most digital preservation infrastructures. The Reference Model for an Open Archival Information System (OAIS) has provided digital preservation initiatives with a standard set of definitions and activities to help guide their development and operations.[1] Specifically, the Reference Model has helped identify what processes and technologies should be present at each stage of submitting, preserving, and disseminating digital objects within a repository. These identifications have proven helpful in solidifying approaches and methodologies within the context of distributed digital preservation.

WHAT IS A DDP NETWORK?

Distributed digital preservation methodologies hold that any responsible preservation system must distribute copies of digital files to geographically dispersed locations. It also must *preserve*, not merely back-up, the files in these different locations. All DDP networks thus share particular characteristics, including that they are comprised of multiple (best practices suggest at least three) preservation sites. Best practices further recommend that these sites be selected with the following principles in mind, all of which are intended to reduce the chances of having any single point of failure that could impact the network:

Sites preserving the same content should not be within a 75-125-mile radius of one another. This helps to ensure that two preservation sites that contain the same content are less likely to suffer catastrophic loss due to disaster scenarios that impact the physical environment.

Preservation sites should be distributed beyond the typical pathways of natural disasters, such as hurricanes, typhoons, and tornadoes. For example, if a network were founded in the Gulf Coast of the U.S., that network would need to establish preservation sites beyond that Gulf Coast region in order to avoid locating all of its preservation copies in a hurricane-prone area.

Preservation sites should be distributed across different power grids. The DDP network should give thought to the power grids that support its infrastructure and avoid placing all copies of any piece of content within the same power grid area.

Preservation sites should be under the control of different systems administrators. For security purposes, all preservation copies should not be accessible by any one person or team of people; instead, control and monitoring of each preservation site should ideally be handled locally by each site in order to ensure that the network's contents are not subject to one point of human-based failure.

Content preserved in disparate sites should be on live media and should be checked on a regular basis for bit-rot and other issues. Because digital content is fragile by nature, degradation may occur at the file level without any external catastrophic circumstances or other easily identifiable triggers. For this reason, content should be evaluated on a regular basis. If evidence suggests that any preservation copy has changed (e.g., through a checksum calculation mismatch), that preservation copy should be compared against the other preservation copies. Once a comparison yields an authoritative version, the degraded file may be replaced with a new copy made from the authoritative file or, in cases where the original is still available for re-ingest, it may be replaced with a new copy made from the original file. This process, like any processes that impact the files, should be recorded permanently within the preservation system to aid in verifying the preservation files' authenticity.

Content should be replicated at least three times in accordance with the principles detailed above. Having at least three replications provides the minimum number of copies necessary to preserve an authoritative file in the face of file degradation/bit rot. When one file no longer matches the others, there are two files (in addition to checksum information and other records about the file) that may be compared to restore an authoritative version as described above. So long as the other principles are followed, it also provides sufficient copies distributed across enough geographically disparate locations to ensure that a copy is highly likely to persist through most foreseeable catastrophic circumstances.

It is worth noting here that cloud-computing servers, which are increasing in popularity, should be subjected to this same set of principles. The servers that enable cloud computing do have a physical location, and that location should be taken into consideration if a cloud environment is used to house one of the preservation copies. If a network chooses to host multiple preservation sites in cloud environments, it must ensure that the geographic location(s) of their servers meet the guidelines detailed above.

Regardless of what technical infrastructure a DDP network adopts, the network will perform three main tasks: content ingest/harvest, content monitoring, and content retrieval. Each is outlined briefly below, but the procedures may vary radically across different technical infrastructures.

Content Ingest

The process of moving content into a DDP network typically takes place in one of two ways. The content may be made available to the network for ingest through the web, or the content may be fed directly into the system through an on-site harvest from a storage medium. Web-based ingest provides a means for all caches to gather content from the same server for preservation without any hardware or storage medium exchanges needed. Given that DDPs are comprised of geographically distant preservation servers, this has proven the most popular choice to date. Whatever the framework, the preservation network must have a way to verify that each copy of the content that has been ingested/harvested matches the original content and thus also each other.

DDP networks, like any preservation system, must take into consideration that digital content experiences a high degree of intentional change. Individual files may be edited and re-saved as the authoritative version. Files may be added to a collection (even a collection that is supposed to be "closed"). DDP networks may want to anticipate such changes by scheduling regular re-ingests of preservation content and/or may want to provide contributing institutions with a clear sense of their own obligations to monitor such updates and additions and share them with the network at regular intervals.

Content Monitoring

Each preservation server must participate in some form(s) of content maintenance and monitoring. This may occur through network-based polling and voting (as in the LOCKSS case described in detail below), through iterative checksum comparisons, or through other comparative mechanisms. The preservation network should provide reports to its contributing institutions on a regular basis that detail the status of the network's servers and the contents that they preserve.

Content Recovery

The DDP network must have a means of retrieving content for a contributing institution when that institution needs the preservation copy to replace a local copy. This should be possible both on a file level or a collection level and the preservation network should be able to deliver an authoritative version of the file in a reasonable amount of time.

The next section of this chapter explores one technical approach, the PLN, as an example of how each of these elements might work.

WHAT IS A PLN?

Within the cultural memory community (comprised of libraries, archives, museums, and other cultural memory entities), many

DDP solutions to date have relied upon the LOCKSS software in a Private LOCKSS Network (PLN) framework. A PLN is a closed group of geographically distributed servers (known as "caches" in LOCKSS terminology) that are configured to run the open source LOCKSS software package. This software makes use of the Internet to connect these caches with each other and with the websites that host content that is contributed to the network for preservation.

In PLNs, every cache has the same rights and responsibilities. There is no lead cache equipped with special powers or features. After a cache is up and running, it can continue to run even if it loses contact with the central server. Such a peer-to-peer technological structure is especially robust against failures. If any cache in the network fails, others can take over. If a cache is corrupted, any other cache in the network can be used to repair it. Since all caches are alike, the work of maintaining the network is truly distributed among all of the partners in the network. This is one of the great strengths of the distributed preservation approach.

The LOCKSS caches of a PLN perform a set of preservation-oriented functions:

- They ingest submitted content and store that content on their local disks.

- They conduct polls, comparing all cached copies of that content to arrive at consensus on the authenticity and accuracy of each cached content unit across the network.

- They repair any content that is deemed corrupt through the network polling process.

- They re-ingest content from its original location (if available) in order to discover new or changed segments, and preserve any modifications alongside the original versions.

- They retrieve and provide a copy of the content that they are preserving to an authorized recipient when called upon to do so (e.g., they can provide a copy of preserved content to the contributor that originally submitted the content if that contributor's own master files become corrupt or are damaged due to natural, technical or human errors. They can also provide a copy of preserved content to repopulate another preservation cache if that cache's contents have been compromised.).

- They provide information to a cache manager tool and the LOCKSS daemon user interface to aid in monitoring and reporting activities.

Each of these activities is described briefly below and elaborated upon in subsequent chapters.

HOW PLNS "KEEP STUFF SAFE": FUNCTIONS AND COMPONENTS

Components of a PLN

PLNs are comprised of a group of distinct and geographically distant preservation sites, each of which runs a server with an installation of the LOCKSS software. Each preservation site serves as a repository for content that is ingested through web-based protocols. PLNs are secure networks and are often run as dark archives that have no public access points. Each server is configured the same way. All are peers with the same rights and responsibilities. Caches in the network are connected to every other cache through the Internet; these servers communicate regularly with one another regarding the content that they are preserving using a TCP protocol called LCAP, which is specifically designed to be difficult to attack.

The LOCKSS team currently recommends that content that is preserved in a LOCKSS network be replicated seven times across the network. If a network contains more than seven preservation sites, not every site needs to store every piece of content; content may be spread across the membership in accordance with the DDP principles outlined above.

Unlike many preservation strategies (especially non-DDP ones), LOCKSS does not rely solely on checksums for preservation. More important in the LOCKSS context is the constant monitoring of all preserved copies. LOCKSS ingests content from the web. At the point of ingest, the LOCKSS software conducts polls across all of the preservation sites that ingest any particular digital file using SHA-1 checksums to ensure that they have identical versions of that file.

The LOCKSS caches repeat the polling/voting cycle on a regular basis after content has been ingested in order to detect anything that has degraded or changed within any given copy of the content. The multiplicity of the content (again, at least seven copies) enables the network to determine, through the polling and voting process, which copies are authoritative and which copies are not (i.e., if bit-rot or other degradation has occurred). Additionally, the caches periodically continue to revisit the site from which the content was originally ingested (as long as that site is available), and wherever they detect changes or additions to this content, the caches ingest and preserve these new or altered files as separate versions. Wherever caches detect content deletion, caches note these as well without deleting previous versions.

This re-ingest feature honors something that most preservation technologies ignore: that one of the benefits *and* liabilities of the digital medium is that once published or produced, a file may be updated or changed by its creator. Such changes happen all the time in our digital environments and often occur without the proper documentation at the local level. By re-ingesting content periodically, LOCKSS is able to preserve more than just a snapshot of a collection at a moment in time—it can preserve the changing form of that collection *over* time.[1] This is key to the philosophical and technical approach of the LOCKSS software and as such, important for prospective PLNs to understand.

Each LOCKSS cache performs core operations for the network, including content ingest, maintenance, monitoring, and recovery. The sections below provide an overview of the tasks the caches and network engage in during each of these operations. Later chapters provide more in-depth coverage of these procedures, particularly Chapter 6: Content Ingest, Monitoring, and Recovery for PLNs, and Chapter 7: Cache and Network Administration for PLNs.

Content Ingest

LOCKSS performs web-based ingests of targeted content that is made available by contributing institutions.[3] Content may be hosted on servers that implement no or few restrictions on access; at the other end of the spectrum, sensitive content may be hosted temporarily on dedicated, access-restricted web servers until enough copies have been ingested into the network.

The preservation process depends upon the contributing institution's (content contributor's) diligence in ensuring that the content slated for ingest is:

1. available at a web-accessible location;
2. organized and documented in ways that promote human understanding of what files are being preserved and the structure in which they are stored; and
3. able to be differentiated structurally from other content hosted by the content contributor that is not going to be preserved in the network.

In order to ingest a collection of digital objects from a particular web site, the LOCKSS software needs information about that collection. In particular, it needs to know that the LOCKSS caches have permission to ingest the content (handled by a manifest page, which is detailed below) and it needs to have a clearly defined set of parameters that the caches should use when ingesting the collection (handled by a plugin, also described below and in Chapter 5: Content Selection, Preparation, and Management).

A manifest page is a web page that gives LOCKSS permission to ingest content from the web server that hosts that manifest page. The content contributor creates this manifest page and ensures that all content that is slated for preservation can be reached by following links from this manifest page. The content contributor may also create a plugin for each collection or may have the LOCKSS team do so, depending upon the PLN's organizational and technical roles and responsibilities (see chapters 3 and 4 for more details). Plugins provide instructions to the LOCKSS software for ingesting digital content into the network by specifying the rules for the web-based crawl, which begins at the manifest page. The manifest page and plugin are the central elements that direct the ingest process for any LOCKSS-based network. There are many resources that assist content contributors in developing these, including documentation produced by LOCKSS and by MetaArchive and made available through their respective websites, www.lockss.org and www.metaarchive.org.

Prior to the ingest of any collection, each plugin must be tested to ensure that it guides the LOCKSS daemons' web crawl such that the initial content ingest covers all intended content URLs and that subsequent re-crawls will discover expected web site changes. Once a plugin works as intended, it is packaged and signed.

Manifest pages are stored on the web server that hosts content and remain under the control of the content provider. The provider should work to ensure that manifest pages remain stable, since plugins rely on their existence and support when they guide LOCKSS caches through their crawl and re-crawl activities.

Plugins are publicized to LOCKSS daemons in a web-hosted directory (or directories) termed a plugin repository. PLNs may capitalize on the infrastructure provided by LOCKSS for this component of their network or may create and maintain plugin repositories for themselves. In the latter scenario, a PLN must have at least one plugin repository, but may also decide to maintain several. Submission of plugins to plugin repositories should be restricted to a few trusted staff members since injecting a broken plugin into a PLN can cause network disruptions.

When a content contributor submits a collection for ingest into the network, the system administrators for each cache in the network must be notified of its existence and trigger the ingest process on their individual caches.[4] When a LOCKSS daemon uses a plugin to conduct an initial harvest, it crawls the content site exhaustively within the boundaries set by the plugin's parameters.

The caches preserve their ingested content in a preservation repository that is defined by the LOCKSS daemon. The preservation repository includes the digital content and metadata describing that content (obtained in various ways as described in Chapters 5 and 6).

Once the content is ingested, someone (usually either the contributing institution, the PLN's central staff, or the Stanford University LOCKSS team) uses the network's monitoring tools, the cache manager and the user interface on each cache, to ensure that the ingest was completed successfully on at least seven caches and that those caches all contain identical copies of each file.

Sometimes a portion of the materials a contributor designates to preserve are not hosted or made available through web-based servers (e.g., collections comprised of copyright-protected works or those that are for internal use). In such circumstances, the content can be temporarily stored on a staging server that is hosted either centrally by the PLN or by the content contributor. In this case, the content that is hosted on the staging server should be ingested once as a "closed" collection, meaning that the caches will harvest this material once and not try to revisit the staging server URL for periodic re-harvests.

Content Preparation

In preparing digital content for ingest, the folder and file structure matters greatly. Consider the difference in the following two scenarios:

1. Contributor A submits an Electronic Theses and Dissertations (ETD) collection to a PLN. The content is stored on a web-accessible server in folders that are organized by year. All embargoed items are stored in folders that are separate from the non-embargoed content. New material for the collection is accrued on an annual basis; the new content is consistently placed in a folder marked with the year of its accession. Metadata describing each ETD object is stored in the same folder. The naming conventions of the filenames for the ETDs and the metadata share recognizable characteristics.

2. Contributor B submits a portion of its digital masters files, which include all of the contributor's digitized image, audio, and video collections. These diverse files are stored in a repository system on a web-accessible server. The files are contained within folders that are organized by collection (where collection is subject-based). Many of the collections continue to grow at an irregular pace that is not tracked in any official manner by the contributor. Multiple constituents have access to these collections and add new files as they become ready. Metadata describing the objects is stored in its own folder, separate from the content, and it bears no recognizable relationship to the content to which it relates. The repository system in which the collections and metadata are stored provides the necessary linkages between the metadata and collections for access purposes.

Both Contributor A and Contributor B can preserve their content using a PLN. Once ingested into a PLN, all of the "bits" of both contributors' files will be preserved equally well by LOCKSS. However, if disaster strikes at these two contributors' sites and they need to retrieve their content from the PLN to repopulate their internal collections, the work they must complete in order to do so will be quite different.

1. Contributor A will find that the clarity of its content's structure serves it well—the content is organized by year, accrues at an annual rate with no changes occurring within files, and includes metadata that has a clear and easy-to-understand relationship to the content it describes. Repopulating the contributor's internal systems will be relatively painless.

2. Contributor B may have a more difficult job. The content that it receives back from the preservation network will look exactly like the data it submitted. So long as the contributor has a way to ingest these objects back into the repository system and reconstruct the links between the content and its metadata, the contributor will encounter no problems. Likewise, if the contributor has documentation that enables it to understand what the preserved content is and where it belongs, the contributor will be able to reconstitute its holdings (and best practices would recommend that Contributor B preserve the software and documentation in the preservation network alongside its content to facilitate this process). However, if the contributor does not have appropriate documentation and processes in place to enable repopulation of its local collections, that contributor may be left with master files and metadata files that are still completely viable but simultaneously troublesome (because, for example, without documentation and mapping information, it may be impossible to know what file 125098.jpg is or where it belongs).

Preservation readiness depends, in other words, on more than simply submitting a collection to a PLN for preservation. The content contributor bears the responsibility of organizing and documenting its collections in appropriate ways prior to their submission.

The content contributor also must structure its collections appropriately for ingest into a PLN. The collections designated for preservation must be broken up into groups of files that the caches can preserve as a unit. These units are termed "Archival Units" (AUs) – they provide the basis for auditing a well-defined unit of content, and allow all caches to know which collections they are preserving, and which other caches have the same collections.

Polling (auditing) an AU requires each cache to compute a (sometimes several) cryptographic hash of all the content in the AU. In order for polls to complete within a reasonable amount of time, this imposes an effective upper bound on the size of an AU. Each AU that is configured on a cache creates a small amount of ongoing overhead in addition to the disk space required to store the content of the AU, so there is an effective upper bound on the number of AUs per cache. These are not hard bounds, and they're changing over time due to improvements in the LOCKSS daemon and increases in CPU speed and disk size. Best practices currently suggest that the average size of AUs should be between 1 GB and 20 GB.

Obviously, not every collection (as conceived by the content contributor) fits this size frame. In order to strike a balance, the contributor may use the manifest page and plugin information to bundle multiple AUs as one collection. For example, contributor C might have 150 GB of content in one digitized newspaper collection. The contributor breaks this collection down into chunks of approximately 10 GB each, using publication years or another characteristic as the parameters. The result is 15 AUs, each of which is defined as one plugin parameter. In the plugin, the contributor describes the 15 AUs that are ready to be harvested. This meets the needs of both the network and the content contributor by bundling the segments of a curated collection for ingest into the network. Please note: as previously mentioned, these are only the *current* guidelines as of 2010, and over time, the system will continue to accommodate more, and larger, AUs.

Maintenance and Monitoring

Ingested content that is actively preserved by a PLN is not merely stored in the distributed network. Rather, the caches work together to preserve this content through a set of polling, voting, and repairing processes. In order to ensure the validity of each copy of the digital object in the network over time, the caches continually poll each other and compare copies to make sure that none of them have changed (due to bit rot or other problems) or disappeared altogether.

During the polling process, the caches compare cryptographic hashes of their respective copies of a particular AU. If the copies are not all the same, the cache or caches with disagreeing files will re-ingest those files from the originating site (the content contributor's site) if they are still available, or will repair their copies from any of the caches that voted in the majority. The re-ingest of repair is noted in the system logs, thus providing a record of every change that takes place to the content over time.

On occasion, an entire cache may crash or disappear due to local conditions. In the event of such a catastrophic system failure the entire cache can be rebuilt in one of two ways:

1. from a fresh ingest of its content from all of the content contributors' originating sites pending their availability, and
2. for whatever content is not available via the originating site, from ingesting content from any of the other caches in the network.

Re-Ingesting Content

As previously mentioned, each LOCKSS cache periodically revisits the site from which it originally ingested a collection in order to check for changes and additions to content. It re-ingests digital objects from the site, not replacing the initial version, but adding a new, dated entry that contains any changed content that it encounters. This is helpful in that most collections change over time. Curators routinely add new content, change files, and otherwise alter the files within their control, and they often do so without notifying others that changes have been made. By periodically re-ingesting materials within each collection, LOCKSS ensures that no valuable content is lost due to changes purposefully made to overall collections.

Likewise, by maintaining the original version as well as the re-ingested versions, LOCKSS ensures that if negative or unintended changes occur within a collection (e.g., files are lost or corrupted or replaced with invalid versions), that the preservation network can return to the content contributor the clean copies that were originally ingested.

Restoring Content

When necessary, any cache may be used to restore content by providing a copy of the content that it is preserving to an authorized recipient. For example, if an institution that has contributed its content to the PLN for preservation finds that its own master files have become corrupted or damaged, that contributor can request a copy of its content from the preservation network. In such a case, any of the caches that hold the content (and as previously noted, each collection should be hosted by at least seven caches) may provide the content contributor with a copy of the digital objects and all other elements of the original ingest process, including submitted metadata and plugin information.

Each cache also provides a user interface (UI) through which authorized users may access the status of the LOCKSS daemon. This user interface enables the administrator of each cache to activate ingest procedures for new collections and to deactivate the preservation of existing content (e.g., if a content contributor no longer wishes for a collection to be preserved). It also provides a means for each administrator to determine that their cache is functioning correctly, to complete performance tuning when necessary, and to troubleshoot any problems that may arise. Importantly, there is no central component that enables changes to be made to local caches at the network level. Instead, a local systems administrator manages each cache. This ensures that the network is not subject to a central point of failure due to human error or an intentional attack on the preservation network.

LOCKSS: BIT-LEVEL PRESERVATION AND FORMAT MIGRATIONS

At its base, the LOCKSS software provides bit-level preservation for digital objects of any file type or format. Bit-level preservation is a foundational element that should be present in any preservation solution. In essence, it ensures that all of the 0's and 1's that make up a given file remain intact, thus maintaining the integrity of the file for later access and use. All PLNs provide bit-level preservation.

LOCKSS also can provide a set of services that work to make the preserved files accessible and usable in the future (including such activities as normalizing and migrating files). LOCKSS first created and tested a format migration strategy applied within the network in 2005.[5] Several PLNs are now beginning to create migration pathways for particular file types that are endangered or obsolete. Once any PLN has created a migration pathway for a file type and implemented it in their LOCKSS environment, it can be used by other PLNs. This provides an additional benefit to the growing community of open source LOCKSS users and developers.

CONCLUSION

All PLNs are not exactly alike. A PLN may consist of a small number of caches, all preserving the same content, or a large number of caches preserving different content. A PLN may keep plugins in a central repository or maintain one repository per contributing institution. A PLN may rely on the Stanford University LOCKSS team to maintain its plugins and archival unit lists or it may maintain these resources itself. There are multiple ways to configure a PLN and, as we will see in the next chapter, there is currently a range of exemplars, each of which has customized its network through the operational and technical decisions it has made according to the needs of its membership.

ENDNOTES

1. Reference Model for an Open Archival Information System (OAIS) http://public.ccsds.org/publications/archive/650x0b1.pdf (last accessed 01-19-2010).
2. LOCKSS can do this successfully for slow-changing sites based on the re-crawl interval that is set for that site. This is not a recommended method of preserving all versions of sites that are changing quickly (i.e., more than monthly).
3. For some PLNs LOCKSS is ingesting and preserving source files.
4. This is not an automated process. Part of the security of the network comes from the lack of centralized (and thus one-point-of-failure) tools for the management of the content preserved across the geographically distributed caches.
5. For more on this format migration work, please see http://www.lockss.org/lockss/How_It_Works#Format_Migration; and http://www.dlib.org/dlib/january05/rosenthal/01rosenthal.html (last accessed 01-19-2010).

Chapter 3: Technical Considerations for PLNs

Beth Nicol (Auburn University)

OVERVIEW

This chapter analyzes the physical and technical organization of various active Private LOCKSS Networks (PLNs). It focuses on the technical decision-making process associated with the creation and maintenance of PLNs, as exemplified through three case studies: the MetaArchive Cooperative, the Alabama Digital Preservation Network (ADPNet), and the Persistent Digital Archives and Library System (PeDALS). For a corollary chapter on content and governance questions, please see Chapter 4: Organizational Considerations.

DEFINING A PRIVATE LOCKSS NETWORK

As mentioned previously in Chapter 2: DDP Architecture, a PLN employs a specialized application of the LOCKSS protocol and function, while using the same software as the original, public LOCKSS network. In accordance with the LOCKSS developer's recommendations for best practices, PLNs performing distributed digital preservation may begin with as few as seven geographically dispersed peer caches. A PLN may collect and archive any type of content, as LOCKSS is format agnostic. Prior to its ingest, this content sometimes requires reorganization in order to be effectively ingested into the private network (please see Chapter 5: Content Selection, Preparation, and Management for more information on this "data wrangling" activity).

As of January 2010, there are at least eight PLNs (see Appendix A for descriptions and contact information) in existence, harvesting a range of digital resources that includes archival image files, government publications, electronic theses and dissertations, audio and video files, and web pages. Most of these networks take a dark archive approach to their preservation, restricting the contents to the network participants for preservation purposes only (i.e., no access is provided to the archive's contents except when needed to restore a content contributor's collections). However, at least one of these networks, the U.S. Government Documents PLN, provides access to cached information in cases of loss of access (as opposed to loss of content) from the primary source.

CREATING A PRIVATE LOCKSS NETWORK

When creating a PLN, the network members must make a number of basic technical infrastructure decisions. The following hardware, software, and network questions should be addressed:

- Which operating system/LOCKSS platform will be used?
- What hardware is required for each member?
- How many participants are required to establish the network?
- How and where will the plugin repository (or repositories) be deployed?
- Where will the title database be located?
- How will plugin development be managed?

Software Installation

The LOCKSS team provides packages for RPM-based Linux systems (such as Red Hat and CentOS[1]) and Solaris systems, which allow users to easily install, configure, and update the software. Most LOCKSS boxes currently employ a third option, a Live CD "appliance" containing a specially configured version of the OpenBSD operating system and the LOCKSS daemon.[2] The LOCKSS team expects to transition from the Live CD to a VMware virtual appliance in the near future.

Two of the three PLNs reviewed here (PeDALS and ADPNet) originally chose to use the OpenBSD CD deployment that is offered by the LOCKSS team The PeDALS project selected the OpenBSD for a number of security related reasons and for a standard configuration across the consortium; ADPNet initially utilized the Live CD installation to facilitate the participation of institutions which might lack the technical expertise required for installing on other UNIX-type platforms. However, because the individual caches have increased to 8TB of storage, ADPNet plans to adopt the package installation for newer caches.[3] The MetaArchive Cooperative elected to create a package installation from the outset. The reason for this decision relates to the MetaArchive network's requirement for an extensible operating system architecture that would allow for modular software development and additional hardware options (more on this architecture below).

Just as all caches in a given PLN have some flexibility with installing the LOCKSS software, they also have some freedom in choosing individual operating systems. Nevertheless, each of the PLNs reviewed here has elected to implement the same operating system on each member cache. This uniformity can greatly simplify both troubleshooting and system upgrades. For more details on this process, please see Chapter 7: Cache and Network Administration for PLNs.

Hardware

The LOCKSS team recommends that users purchase low-cost hardware from small vendors, particularly if they plan to use the OpenBSD deployment. Smaller vendors are slower to turn over their inventory, and thus tend to have slightly older chips installed in their machines—a benefit, because the open source community is more likely to have built drivers for slightly older chips (of particular interest here is the OpenBSD team).

There are two options for PLN hardware selection:

1. all caches use identical hardware; or
2. members agree to a set of minimum requirements.

With the second option, requirements should specify a minimum amount of RAM (memory), a minimum processor speed, and a minimum amount of disk space for content storage. To date, most PLNs have found that using identical hardware is the most manageable way to operate to a network. This approach simplifies troubleshooting, as well as the ingest of archival units, due to compatible CPUs and disk configurations.

Regardless of what hardware is selected by a network or an individual cache, best practices strongly recommend that each server be designated specifically as a preservation cache and not perform any other function.

Network Components

Plugins describe the characteristics of collections or similarly structured groups of collections for ingest and preservation. Plugins are stored in one or more plugin repositories and are made web accessible to the individual caches in the network. There are four models for how to host a plugin repository:

1. as a single repository hosted by the LOCKSS team;
2. as a single repository on one of the local PLN caches;
3. as one of multiple plugin repositories for a single network; or
4. as a central component hosted on an administrative server that is not a LOCKSS cache.

ADPNet and COPPUL have thus far relied on the model of a single repository hosted by the LOCKSS team. The Cooperative has experimented with the last three of these methods. Based on its experience, a single repository appears to meet the technical and administrative needs of a network, and is easier to maintain than multiple repositories hosted across the network. Currently, the MetaArchive network is running a central plugin repository hosted by a cloud computing service to ensure equal access to all contributing members. In the PeDALS project, each system is self-contained and maintains a central plugin repository and title database hosted on a separate server, which is available only within the individual member's network.

Plugins must be developed by, or in close cooperation with, the content contributor. A content contributor in a PLN is a member site that prepares collections for harvesting. Unlike the public LOCKSS network, most content contributors within a PLN also serve as content harvesters.

Another component of the network is the title database, which connects plugins to their corresponding archival units (AUs), making the AUs selectable for harvest. The LOCKSS team at the Stanford University Libraries hosts the title database for those PLNs that rely upon the LOCKSS team for network management and infrastructure. A title database can be implemented separately from the main LOCKSS unit for those PLNs that desire greater independence and flexibility. For further discussion on these two implementations see Chapter 7: Cache and Network Administration for PLNs.

CASE STUDIES

The previous section provided an overview of some of the technical decision-making that should take place when establishing a PLN. This section further illustrates these decisions through brief case studies of three well-established PLNs: the MetaArchive Cooperative, the Alabama Digital Preservation Network (ADPNet), and the Persistent Digital Archives and Library System (PeDALS).

Each of these projects shares the same broad goal of preserving digital information; however, each has approached the establishment of its PLN in distinct ways.

The MetaArchive Cooperative

The MetaArchive Cooperative worked with the LOCKSS team to develop the first PLN. With funding from the Library of Congress's National Digital Information and Infrastructure Preservation Program (NDIIPP), the Cooperative sought in 2004 to extend the use of LOCKSS to preserve digital content beyond the initially targeted realm of electronic journals. This funding provided the original technical support resources for the Cooperative's network. The pioneering nature of the Cooperative venture drove many early decisions, including the choice of hardware and operating system, as well as decisions concerning the management and hosting of the plugin repositories and title database. It also determined the scope of the materials preserved in the network, not to mention governance and sustainability issues.

From its earliest days, the MetaArchive Cooperative planned to create modular tools to work with LOCKSS for cataloging and monitoring purposes. For this reason, the Cooperative opted to use an extensible operating system architecture that would allow for such software development. The Cooperative elected to use RedHat Linux and to install the RedHat Package Manager (RPM) version of the LOCKSS daemon. This installation enabled more customization than the CD install and increased the Cooperative's hardware options. This decision also resulted in a need for more technical expertise and support at each site to set up and maintain a cache. It has enabled the Cooperative to customize its PLN, including through enhancing the security of its network.

The Cooperative provides a flexible hardware option for its members by setting a minimum set of requirements for processor speed and disk space. Simultaneously, the Cooperative negotiates a hardware configuration. All caches (except for the Library of Congress) have thus far deployed identical hardware (with a new hardware configuration selected for new members and replacement caches on a regular basis). Initially, the MetaArchive network also included a Dell AX100 RAID-5 storage array as the location for cached files on each server for further redundancy. In retrospect, this arrangement required more effort in set-up and maintenance, and proved to be unnecessarily redundant in the context of the LOCKSS-based network itself.

The Cooperative began in 2004 with multiple plugin repositories, each deployed on the same host as a LOCKSS daemon. This environment required an increased commitment to supporting technical manpower—each contributing member was required not only to create the plugins, but also to perform the programming and system administration procedures necessary to deploy them. Some of the participating members experienced problems in packaging, signing, and deploying plugins using these distributed repositories. In order to decrease such problems and the overhead incurred by member institutions, the Cooperative consolidated the plugin repositories into one repository in 2008. Now, the central staff of the Cooperative manages that repository so that the technical expertise required to prepare plugins for deployment is no longer required at every site.

The Cooperative found that the title database alone was insufficient for maintaining and disseminating the relevant metadata about collections that its membership desired. An additional application, the conspectus database, was developed by the Cooperative to provide a method for storing collection-level metadata. This expanded metadata schema provides information about Cooperative collections (not individual digital items). In addition to collection level metadata, the conspectus database includes plugin information required for harvesting a collection, and allows for the definition of multiple AUs within a collection.

The Cooperative has also partnered with the LOCKSS team to develop a cache manager that provides extended monitoring, reporting, and management of the network as a whole.

The MetaArchive network is a dark archive with no public interface, and communication between caches is secure. In all cases, the content preserved in the network is available only to the contributing institution that owns that content; in some cases, the contributing institution may also make that content available to the public through a separate repository system. The Cooperative governs its dark archive through both technical strategies and legal agreements that are made between all members.

In this dark archive context, firewall rules are an ongoing concern. For example, to protect shared resources, firewalls must be continually updated as caches join or leave the network. Depending on how institutions assign responsibility for configuring firewalls, this task can prove to be either easy or problematic. For further discussion of configuring firewalls and ports for a PLN see Chapter 7: Cache and Network Administration for PLNs.

The Alabama Digital Preservation Network (ADPNet)

The Alabama Digital Preservation Network (ADPNet) is a self-sustaining statewide digital preservation system with a low threshold of technical expertise and financial requirements for participation. ADPNet's start-up was supported by an IMLS National Leadership Grant from 2006 through 2008. The network is now self-sustaining and is part of the Network of Alabama Academic Libraries (NAAL), a state agency in Montgomery. As with other PLNs, content selection is limited by available storage. ADPNet currently has seven institutional members, including the state archives. Membership is open to all libraries and cultural heritage organizations in Alabama and does not require a separate membership fee. ADPNet's goal is to continue to function and grow as a self-sustaining network without requiring external funding or membership fees.

Initially, ADPNet chose to use the Live CD install of the LOCKSS platform. In this scenario, the OpenBSD operating system ran from a CD-ROM, with the configuration parameters stored on write-protected media. This type of installation and configuration is similar to installing a cache on the public LOCKSS network. In ADPNet's case, the write-protected media was either a flash drive or a CD containing the LOCKSS software. The flash drive required a write-protect option so that the configuration data could not be accidentally overwritten and to provide the appropriate level of security. With this method, no expertise is required to install and configure the operating system. Also, security is inherent, because the operating system loads and runs from non-writable media. It does somewhat restrict hardware options, because it requires off the shelf components, which are likely not the newest options, and may no longer be supported. A benefit of this type of installation is the automatic updates it provides to versions of the LOCKSS daemon.

As the content for the network has grown, and the storage capacity requirements for network caches have increased, ADPNet has begun to use the package install of the LOCKSS daemon for newer caches. The LOCKSS team advised this change for performance reasons when ADPNet caches reached 8TB of storage.

To further facilitate the network set-up, ADPNet decided to use identical hardware for each cache, using inexpensive servers with specifications suggested by LOCKSS. As it did with the operating system, the ADPNet chose no-frills hardware. ADPNet continues to encourage the use of identical hardware when possible.

The ADPNet receives network-level technical support from the Stanford University LOCKSS team who hosts both its plugin repository and title database. This configuration removes ADPNet's local need for technical expertise to deploy plugins and generate the title database. Plugin development is handled at each participating site; the final testing and deployment are handled by the LOCKSS team.

Like the MetaArchive Cooperative, ADPNet has found a need for better tracking and descriptions of collections. ADPNet is interested in deploying the conspectus database developed by the Cooperative to facilitate better tracking of collections within the network.

ADPNet is also a dark archive with no planned public interface. Currently, the ADPNet is primarily ingesting content on open web servers. However, some collections that are being ingested reside on access restricted or closed web servers.

The Persistent Digital Archive and Library System (PeDALS)

PeDALS is a grant-funded research project of the Arizona State Library, Archives and Public Records, in collaboration with the Alabama Department of Archives and History, the State Library and Archives of Florida, the New Mexico State Records Center and Archives, the New York State Archives, the New York State Library, the South Carolina Department of Archives and History, the South Carolina State Library, and the Wisconsin Historical Society. PeDALS was founded to investigate a rationale and software for an automated, integrated workflow for curating large quantities of digital records and publications, in addition to developing a community of shared practice. PeDALS made start-up decisions that sought to accomplish these goals while removing barriers to participation by keeping technology costs as low as possible. Funding sources are the Institute for Museum and Library Services (IMLS) Library Services and Technology Act (LSTA) and the Library of Congress National Digital Information Infrastructure and Preservation Program (NDIIPP).

LOCKSS is only one component of PeDALS. The PeDALS PLN functions as a dark archive that is protected through a set of robust firewalls. This implementation insures the integrity of the records being stored in the network. The PLN component of PeDALS requires strong security due to the sensitive nature of much of the archived material. Since the PeDALS members wished to implement a system that took almost no local system administration expertise, they choose to install the LOCKSS software from the Live CD.

The PeDALS Core Metadata Standard[4] includes elements common to government records, including administrative, technical, and descriptive information. For items such as marriage certificates, the ingested object consists of a PDF file plus a metadata file in XML format.[5] The New Zealand Metadata Extractor[6] is used to generate technical preservation metadata, such as size, format, mime-type, and application.

PeDALS differs from the MetaArchive network and the ADPNet in that PeDALS is a group of isolated networks, as opposed to a single network with several members. Each contributing member is expected to support an entire cluster of seven systems to preserve its own content. Some servers in the PLN remain in the home state, while others are distributed to at least two other members. This distribution provides the geographical diversity of the networks.

In PeDALS, a distinction is made between restricted and non-restricted content, and the open, non-restricted content will be available to users over the Internet. However, the public interface for non-restricted content will not access the LOCKSS cached content, but will operate from copies of the objects exported from the LOCKSS caches.

ADDITIONAL CONSIDERATIONS

PLNs are evolving entities. Since its inception, the MetaArchive Cooperative has explored and utilized various hardware configurations, as well as various configurations for the plugin repositories and the title database. The ADPNet began with identical hardware and a Live CD install, and has progressed to encompass both non-uniform hardware and the RPM installation. PeDALS has expanded beyond its original membership. The technical considerations described in this chapter are a good starting point, but cannot reflect the evolving nature of PLNs. For more information about individual PLNs, please see the contact information in Appendix A.

ENDNOTES

1. http://www.lockss.org/lockss/Installing_LOCKSS_on_Linux (last accessed 1-28-2010).
2. David S. H. Rosenthal. "A Digital Preservation Network Appliance Based on OpenBSD". In *Proceedings of BSDcon 2003*, San Mateo, CA, USA, September 2003.
3. Once the amount of content in a cache reaches a certain high quantity, the Live CD (OpenBSD) implementation becomes less optimal as Linux systems can handle large disk arrays more efficiently than OpenBSD.
4. PeDALS Core Metadata Dictionary (Draft 3.0, 14 November 2009),. http://pedalspreservation.org/Metadata/PeDALS_Core_draft_3_2009-11-14.pdf (last accessed 1-22-2010).
5. The received metadata fields for this item type includes bride, groom, marriage date, recording date, and license number. The bride's and groom's names are stored as "Parties to the record," a name-value element that associates access points with their roles (name=bride, value="Smith, Jane", name=groom, value= "Doe, Joe"). Similarly, the marriage are recording dates are stored in as ItemDate, another name-value pair that associated a specific date and the nature of that date in the context of the record. In some instances, metadata can be supplied by default rule. For example, the digital image of the record includes the city. Because all records in this series come from the same county, the rule can supply the county name for the location, providing a minimum amount of information.
6. http://meta-extractor.sourceforge.net/ (last accessed 1-24-2010).

Chapter 4: Organizational Considerations

Tyler O. Walters (Georgia Tech)

OVERVIEW

When institutions join together to manage the technologies that comprise a Private LOCKSS Network (PLN), they must first give thought to the organizational framework in which they will function. This distributed digital preservation (DDP) solution depends upon a collaborating set of institutions agreeing to preserve each other's content. The findings of this chapter, though tied to PLN case studies, are extensible to other DDP solutions as well. There are many organizational considerations involved in establishing a collaborative association of institutions for the purpose of long-term preservation. Among these considerations are issues relating to contributor roles and responsibilities, governance structures, agreements between contributors, and member staffing requirements. This chapter covers these topics, using existing PLNs as examples. For information regarding the technical decisions new networks must make, please see Chapter 3: Technical Considerations for PLNs.

ROLES AND RESPONSIBILITIES

Overview

When cultural memory organizations collaborate, they tend to do so through short-term grant funding opportunities, which usually last around two to three years. Collaborating on digital preservation demands a different type of relationship, one with long-term aims and a formal, sustainable infrastructure. PLNs have been experimenting with such infrastructure development since 2004. This section uses three PLN governance models to examine the roles and responsibilities incurred by content contributors in a distributed digital preservation network.

Governance and Membership Models

Many different models of governance can be applied to a PLN, as well as to DDP networks. The main reasons for establishing a governance structure are to:

1. establish a decision-making body and procedures for decision-making;
2. provide a means for communication and collaborative engagement across content contributors (particularly important as these are geographically distributed solutions); and
3. implement proper management oversight – strategic, fiscal, and operational.

There are at least two governance models emerging from the establishment of PLNs and other solutions (see also Provan and Kenis, 2007).[1] These models are:

1. The Network Administrative Model (e.g., the Educopia Institute as host organization and management entity for the MetaArchive Cooperative; and CLOCKSS Archive as the community host organization for the CLOCKSS PLN); and
2. The Lead Organization Model (e.g., the PeDALS PLN led by the Arizona State Library and Archives; the SRB-based Chronopolis Network led by the University of California at San Diego; and the MetaArchive Cooperative as it was led by Emory University during 2004-2007).

Each governance structure must address essential questions regarding the management of the members, services, and functions of a PLN. Among these questions are:

- Who can join the organization?
- What are the contributor responsibilities?
- Who leads the group? Is it led by a peer contributor model or by a host institution model?
- What are the responsibilities of the lead entity?
- How does a contributor leave the PLN?
- What happens to a former contributor's content?

- Can a contributor's membership be revoked? Under what circumstances?
- How is the governing body organized?
- What type of accord governs the group? For example, is there an organizational charter, memorandum of understanding, membership agreement, service-level agreement, informal (spoken) understanding, etc.?

COLLABORATIVE NETWORK FORMATION

As described in Chapter 2: DDP Architecture, the field of distributed digital preservation is still in its infancy. As such, there are very few examples of successful DDP models to date. This section focuses on three collaborative network models that are available to new and existing DDP networks and provides brief notes on the benefits and drawbacks of each. It also shares several case studies that illustrate one of those collaborative network models. These case studies all use one DDP technical framework that has been deployed repeatedly within the library community to great effect to date: the PLN. These organizational approaches, however, are extensible to different DDP network forms and other collaborative ventures as well.

The Participant Model of governance intends to address and directly alleviate the potential imbalance that can arise between a lead member (discussed below) and other peers in the network. Peer institutions participate equally in the management of the PLN, usually through executive board and committee structures. While providing a structure for many opportunities to participate in the network's management at equal levels, this model does not directly provide for a central mechanism to collect, manage, and implement staff, technology, and finances. It may leave the network vulnerable to contesting viewpoints unless there is a clear leader designated to make important decisions.

The Network Administrative Model is similar to the participant model in that all peers in the network operate as equals. However, it also includes a centralized management and administrative structure providing clear leadership channels. It attempts to bring together the benefits of the two other models: high levels of governance participation from multiple institutions and central administrative and management services. However, the creation and maintenance of the central apparatus introduces additional costs that must be shouldered by the member institutions.

The Lead Organization Model provides centralized resources and management for the preservation network through one of its members taking on central responsibilities. One lead institution typically uses its existing structures to supply and manage staff, technology, and finances, as well as to conduct managerial decision-making. This model implements centralized management and operations that are run by one of the peer institutions of the network. Benefits include a clear sense of leadership and responsibility as well as a lower overhead for the network's operations, as the lead institution provides these basic management structures. Drawbacks include the network's dependence on that peer institution's continued support for the network's operations and a potential sense of imbalance among the member institutions.

The governance structure employed by a network has an enormous impact on its stability and sustainability. Particularly in the emerging field of digital preservation, institutions must ensure that they are creating long-term strategies, not just technically, but also organizationally.

Each of the selected case studies featured below employs the "Network Administrative Model," but as the case studies show, this model may look quite different in practice depending on the type of administrative apparatus that is used or implemented for governance.

CASE STUDIES

As mentioned previously, the case studies selected for this chapter use the PLN as their technical framework. All PLNs share certain organizational characteristics, regardless of their governance structure. For example, all PLN content contributors that run LOCKSS caches are required to join and maintain membership in the LOCKSS Alliance, whose function is to develop and maintain the LOCKSS software. All PLNs are currently composed of at least seven members managing a minimum of one cache each, as the LOCKSS team has recommended that at least seven copies of each content submission be maintained in the PLN environment. Also, no one cache administrator has full access rights to all the servers in a network, thus further ensuring security and avoiding a single point of failure in regards to the risk of human error from one person with access to all systems.

Despite these similarities, each PLN may look and function quite differently, according to the specific organizational decisions each makes, as illustrated in the case studies below.

MetaArchive Cooperative

The Cooperative began in 2004 as an inter-institutional collaboration of six universities plus the Library of Congress. It formed its own non-profit management organization, the Educopia Institute, in 2006. Educopia provides administrative, fiscal, and strategic oversight to the Cooperative in cooperation with the MetaArchive Steering Committee, which guides and oversees the various subject- and genre-based networks and programs of the Cooperative. Within the Cooperative, functional committees composed of representatives from the content contributors communicate and advance the Cooperative's work. These committees are the Content Committee, Preservation Committee, and Technical Committee.

The MetaArchive is a true cooperative of contributing institutions. It depends on its members to participate significantly in its digital preservation activities and, therefore, to shoulder the responsibilities of preservation. It is not a vendor-like business model where services are completely operated by the vendor and provided to customer institutions for a fee.

Management organizations like Educopia create a level playing field between content contributors, such that no one contributor plays a dominant role based on its organization size or level of resources. Each content contributor within the same membership category pays the same annual dues to Educopia and has the same number of representatives on MetaArchive committees. A minimum of two Cooperative members serve on the Educopia board at all times. The Cooperative operates under a charter, which details the mission of the Cooperative, membership costs, benefits and responsibilities, organization and governance, and services and operations. A related membership agreement documents the roles and responsibilities of each content contributor and the Cooperative to one another and serves as the legally binding contract for member organizations.

ADPNet

ADPNET was formed under the auspices of the Network of Alabama Academic Libraries (NAAL), which also provides it with an administrative home. The ADPNet is composed of Alabama-based academic institutions, state agencies, and cultural heritage organizations. This initiative is managed directly by an elected steering committee with representatives from each of the seven member organizations, which are college and university libraries and one state cultural agency. The steering committee works in consultation with the members and makes all decisions for ADPNet, which is managed by a governance document specific to the network. This model is an economic, efficient, and simple approach to developing an organization to manage a PLN within an existing administrative apparatus. There are no direct membership fees, since the network's modest administrative costs are shouldered jointly by the NAAL and the member organizations and are covered through the state. Members that host caches for the network are required to join the LOCKSS Alliance. This approach uses the Network Administrative Model, drawing upon features of the participant model for governance. ADPNet organized itself with NAAL as its administrative entity; however NAAL does not participate directly in ADPNet management – ADPNet's steering committee performs this function.

CLOCKSS

Controlled LOCKSS is composed of governing and advisory member libraries and publishers who have joined together to preserve selected publishers' scholarly content. This initiative is managed directly by a representative board composed of individuals from each of the governing libraries and publishers. The board receives input from an advisory council consisting of individuals from supporting member libraries and publishers. The governing and supporting members pay fees to support CLOCKSS. In exchange, members have a seat on the Board or Advisory Council, which manages the CLOCKSS organization. Fees to CLOCKSS also subsidize the purchase of the caches present on the CLOCKSS network. The CLOCKSS Board makes all decisions related to the technical and organizational network. This model illustrates how revenue-generating producers of information (publishers) and universities (via their libraries) can act together on behalf of the public good, by building an organizational structure to preserve and provide free, open access to scholarly content when it is not available from any publisher.

The CLOCKSS organization is managed by a governance document.

Organizational Structures: Impact

Beyond the Model deployed, the details of the organizational structure chosen by a network has a measureable impact on the decisions it makes regarding the roles and responsibilities of its content contributors. This section draws upon a comparison across the case studies to illustrate the effects of the organizational structure on the administrative design of a PLN.

Member Roles

The roles of content contributors differ from one PLN to another.

In the case of the MetaArchive Cooperative, there are currently two membership levels:

- Sustaining Member – develops network technology, operates a preservation cache, contributes to and helps to direct the development of the MetaArchive network through representation on the MetaArchive Steering Committee; and
- Preservation Member – operates a preservation cache, and contributes and ingests content (i.e., does not develop network technology or participate on the Steering Committee).

In the case of the ADPNet, all contributors supply network content, participate in network governance, and rely on the Stanford LOCKSS team for technical support. In PeDALS, each contributor supplies content only within its own network, and each network is distributed across member sites.

Content and Mission

Any new multi-institutional DDP or PLN must define the network's sphere of activity. This decision largely revolves around the content coverage of the preservation network. This scope could be based on subject matter (e.g., southern culture), chronology, geography (e.g. Alabama), item type (e.g., electronic theses and dissertations, social science data), or institution or organization (e.g., a network open to any digital holdings from the network's content contributors, such as CLOCKSS). Once defined, the content coverage forms the cornerstone of the new PLN's mission, which is preserving the identified body of digital content and restoring it to content contributors when called upon to do so. See Chapter 5: Content Selection, Preparation, and Management for more details about managing issues of content and mission.

Institutional Participation

Concurrent with determining the content coverage and mission of the PLN, contributing institutions must also identify additional institutions and organizations for potential participation. The best approach to date has included outreach to entities that have a strong stake in using, preserving, and providing access to the preserved content. Prospective content contributors should also be able to bring certain attributes to the PLN. These resources may include staff and expertise, facilities, hardware and software, or additional financial resources. These stakeholders will, in turn, have the proper motivation and incentive to support the PLN, as they have a vested interest in the content.

In the case of the MetaArchive Cooperative's Southern Digital Culture archive, content contributors include archives that have a mission to document and preserve southern culture and history. Similarly, in the MetaArchive Cooperative's ETD archive (co-hosted by the Networked Digital Library of Theses and Dissertations, NDLTD), content contributors include academic institutions with a mission to preserve a genre-based electronic theses and dissertations collection. In the CLOCKSS network, contributing institutions consist of publishing companies producing scholarly content and libraries interested in preserving and providing continuing access to that content. In the ADPNet, contributing institutions include college and university libraries and a state agency, all with a mission to preserve cultural materials relating to Alabama.

Services and Benefits

Each PLN should describe the benefits of member participation and the services rendered in an easily identifiable set of documentation. The benefits to a contributing institution generally include the preservation of digital materials and the services to restore them to the contributor.

- The MetaArchive Cooperative's services are rendered to members as well as non-members. Non-members can receive digital preservation network consultation and assistance, digital collection disaster assistance, educational programs, and general

LOCKSS services. Also, the scholarly community and the general public benefit from the Cooperative's work to preserve materials that are often used by researchers.

- The ADPNet's benefits are to contributing institutions; however, by extension, the general public benefits as well, because contributing institutions are tasked with preserving Alabama's cultural materials for the public. ADPNet also shares information with and consults with other institutions that are interested in DDP.

- CLOCKSS benefits scholars worldwide, as a *trigger event* makes CLOCKSS-preserved content open and free to the public. A trigger event occurs when archived content is not available from any publisher. Institutions who support the archive actively govern the archive, including participating in setting fees, strategic direction, and determining when content should be copied from the archive and made freely available to all.

STAFFING RESOURCES AND FUNCTIONS

A key difference between a distributed digital preservation model (particularly the PLN) and other preservation solutions is that the DDP often depends on a distributed staffing model to accomplish its goals. In the PLN case, each PLN relies on its contributing members to fulfill particular roles that may be filled by any number of staff members. Central functions for a PLN include network program management and systems. Distributed functions for each contributing institution that runs a cache include cache administration, collections management, and content preparation for harvest. Other roles at the network level include legal advisors with knowledge of copyright, intellectual property, contract law, and in some cases, non-profit organization.

Some of the central positions present in existing PLN organizations are described briefly below; however, PLNs also depend upon local systems administrators to run caches. In some PLNs, local programmers write plugins to enable content ingest, and local curators to make selection and documentation decisions about the preservation process.

Central staffing needs of a PLN may include:

Program Manager:

The Program Manager coordinates communication among content contributors and their distributed staff. The Program Manager writes and maintains documentation about the organization and its technical infrastructure, organizes meetings for members, ensures that members are trained and integrated into the network, administers the PLN's communication tools, monitors and reports on the network, publicizes the work and achievements of the network, recruits new members, and identifies grant opportunities to support the network's development. Some PLNs have hired staff to do this, such as the MetaArchive. Other PLNs share this responsibility on a rotating basis, for example ADPNet.

Systems Administrator:

The Systems Administrator ensures the reliable operation of a network of caches created to mutually preserve digital content. The Systems Administrator installs, configures, provisions, and maintains secure Linux installations and Apache http servers; adapts, configures, tests, and documents kickstarts for network members; monitors the network, and maintains the central tools for the PLN (including the title database and the cache manager). The Systems Administrator also helps to train the cache administrators at contributing institution sites.

Software Engineer

The Software Engineer provides technical leadership for the development of modular software components for use in the PLN, including network status monitoring tools, reporting systems, and browsing interfaces for content deposited in the system. The Software Engineer also provides training and assistance to content contributors as they prepare their content for harvest into the network, including writing manifest pages, writing and testing plugins, and advising on content organization strategies.

Member site staffing (or roles) may include the following:

Collections Manager/Archivist:

The Collections Manager leads the selection of a contributor's digital collections for preservation in the PLN. This person is responsible for the conspectus database metadata, the LOCKSS title database metadata, and other related database content. The Collections Manager may oversee the work of the plugin programmer and data wrangler to prepare digital collection content for harvest and may be involved in other preparatory steps to create and harvest digital collections.

Cache Administrator:

Each local Cache Administrator establishes, maintains, and monitors the individual cache, monitors the cache's relationship to the overall network, coordinates with the other systems administrators, and conducts network testing as needed.

Ingest Module (Plugin) Programmer:

The Plugin Programmer creates plugins and manifest pages for all collections submitted for harvest by an individual content contributor. Many PLNs currently rely on the Stanford LOCKSS team to create plugins and to advise each PLN member where to store their permission statements and how to organize their content.

Data Wrangler:

The Data Wrangler readies a contributor's content for harvest and prepares conspectus entries. This individual works with the transfer of large amounts of digital information and collections, typically from temporary storage media, web servers, or repositories, to the PLN.

Other Staff Considerations

Legal Counsel:

Legal counsel is helpful (if not imperative) for creating and reviewing key organizational documents that establish the PLN organization and its memberships. Such counsel is also needed to review intellectual property-related polices regarding content ownership during both the network ingest phase, as well as the retrieval phase for preserved content. These policies affect the preservation rights and eventual usage and distribution of content from the PLN. For further discussion on issues related to copyright in the context of PLNs see Chapter 8: Copyright Practice in the DDP.

Other Resources:

Financial and technical support is necessary for geographically dispersed PLN contributing institutions to travel to PLN meetings and to communicate over distance via phone, Internet technologies, or other forms of telecommunication.

CONCLUSION

Institutions and organizations interested in utilizing distributed digital preservation should scrutinize existing organizational approaches and consider closely how they can best leverage them. Issues such as governance, institutional roles and responsibilities, network infrastructure, as well as considerations of staffing and functional support should be taken into account. The decision-making criteria should be based on identifying which aspects of these issues best serve the contributing institution, as well as how the institution can optimize its support and contributions to the larger network of institutions in the PLN.

ENDNOTES

1. Provan, Keith G. and Patrick Kenis, "Modes of Network Governance: Structure, Management, and Effectiveness," *Journal of Public Administration Research and Theory* 15/1, 2007 http://jpart.oxfordjournals.org/cgi/content/full/mum015v1 (last accessed 01-10-2010).

Chapter 5: Content Selection, Preparation, and Management

Gail McMillan (Virginia Tech)

Rachel Howard (University of Louisville)

OVERVIEW

This chapter covers recommendations for policies regarding the selection, preparation, and management of digital content preserved in a Private LOCKSS Network (PLN). It provides best practices for organizing and data wrangling collections of both scanned and born-digital materials. These best practices may also be applied more broadly to many distributed digital preservation (DDP) initiatives, as the fundamental principles of collection organization will be similar across such approaches. These policies and best practices may affect existing digital collections as well as the planning of future digital projects. This chapter also describes some of the specific techniques and technical steps that participants in a PLN will need to follow as they ready their collections for ingest by the LOCKSS software. For further best practices regarding ingest, monitoring, and recovery of digital content preserved in a PLN, please see Chapter 6.

CONTENT SELECTION

We cannot preserve everything digital, nor would it be particularly useful to do so. Digital content, just like print and object-based content needs to be identified, collected, organized, prioritized, and preserved.

A key difference between traditional and digital preservation is that digital preservation needs to start early enough in the digital object lifecycle for it to be viable. While a brittle book or sticky acetate tape may be salvaged for reformatting after the damage has begun, even slightly corrupted digital materials are not as easily rescued. They cannot be subject to benign neglect, or be created and then ignored for decades, since their formats, software, and/or hardware may degrade, or become obsolete and their storage locations obscured thus rendering them inaccessible over a relatively short span of time.

An early step toward preserving digital files is to identify one or more experts within an institution to determine exactly what content should be preserved. These experts are generally librarians, archivists, curators, and the like who are knowledgeable about digital formats, issues of scope, copyright status, as well as the risk factors associated with digital archive content. These elements are discussed in greater detail below.

Formats

Decisions regarding what formats to preserve may be made at the network level or the local level for any distributed digital preservation solution. Some solutions will dictate format as part of the software itself and others will preserve any file regardless of its format. For example, the LOCKSS software is format agnostic, meaning it will accept any computer file in any format. Therefore, contributing institutions (or content contributors) that are planning to host a PLN in particular have two options regarding the file formats they accept and ingest for preservation. They may establish criteria for participation that are format-based, or they may leave the decision about what formats are worth preserving to individual content contributors in their network. In other words, PLNs can be as broad or as narrow as they choose when establishing format-based criteria for participation, and need not be constrained by format decisions made locally at their contributor sites.

Whether the DDP network (PLN or otherwise) or the content contributor makes the decision regarding what formats it will preserve, there are emerging community-wide best practices that can guide the decision-making process. The main consideration, of course, is that some formats will be more accessible than others in the long term. Rare or esoteric formats may require more ongoing investments by content contributors in order to maintain their viability over long periods of time. Bit-level preservation should ensure that content is being responsibly preserved and managed in the interim of any major advances in format validation and migration. One such resource that is advancing, and may prove useful to DDP networks in the future, is the Unified Digital Formats Registry (UDFR), which is an international alliance that is creating a format registry to identify formats approaching obsolescence, and new successor formats that may be viable for migration.[1]

There are also several publications currently recognized in the library community as excellent guides for evaluating file formats for long-term preservation. Two in particular provide specific criteria for determining the durability of any file format, recognizing that the future will call for migration to new formats, emulation of current software on future computers, or both.

The first of these guides is *Sustainability of Digital Formats: Planning for Library of Congress Collections*.[2] This publication describes seven factors that influence the feasibility and cost of preserving any particular file format. In addition, the article discusses quality and functionality factors to consider, as well as the need to find a balance between best practices and the realities of donated digital objects that can be quite varied.

The second publication is an article from the National Library of the Netherlands (NLN) that identifies seven sustainability criteria. Titled "Evaluating File Formats for Long-term Preservation," the article provides quantifiable measures for each criterion, acknowledging that pragmatically not all criteria are equally important.[3] The weights that NLN applies can be adjusted for other organizations, as they are based on a combination of local policy, digital preservation literature, and common sense.

These two examples of file format selection guidelines for preservation recognize that there are a number of potentially competing sustainability factors that must be weighed on an individualized basis. They help to provide readers with an awareness of the range of file formats, some of which can be virtually guaranteed to be sustainable, some of which are likely to be sustainable, and some of which the level of sustainability is as yet unknown.

Another format consideration for DDP networks and their contributors is the determination of what constitutes a master file, and whether this master version is the only one to preserve, or if it is also desirable to preserve derivative (i.e., access) versions. Some consider the master file to be the original scan, original video-capture, or the first digital form of an object (born-digital or digitized). Others consider it to be the richest version of the file that is in use; for example, the master file of a scanned book page would not necessarily be the raw scan or capture, but rather the uncompressed file that has been cropped, rotated, and color corrected for production purposes. Some consider the version of the digital object that best represents the original content to be the best version to preserve. Organizations such as the Digital Library Federation (DLF) have developed standards for elements of a digital master registry, such as the recommendations found in their "Registry of Digital Masters Record Creation Guidelines."[4] Each content contributor at the network level should carefully consider future use scenarios and make preservation decisions accordingly. It is highly recommended to preserve both master and derivative files so long as the network has (and the contributing institution can afford) adequate storage space.

Scope

The decision to form a DDP network is often made based on a shared collection focus that is common among the prospective contributing institutions. The common areas may fall under any number of criteria, such as topical content (e.g., the MetaArchive of Southern Digital Culture archive), genre (e.g., electronic theses and dissertations, or U.S. government documents), format (e.g., data sets), and even location (e.g., statewide initiatives such as the Alabama Digital Preservation Network). This common ground among disparate content contributors may be determined through informal networking or through the analysis of collection data, such as that collected by survey (e.g., the "Electronic Theses and Dissertations Preservation Survey" conducted by the MetaArchive Cooperative in collaboration with the Networked Digital Library of Theses and Dissertations).[5]

It may seem obvious to preserve subject-specific digital materials in a DDP when all the content contributors have holdings in similar fields. However, if inclusiveness is a priority, contributing institutions can select shared subject matter, but define the scope very broadly to include materials that may not be immediately obvious. For example, the MetaArchive Cooperative decided to define its focus of Southern heritage very broadly for its Southern Digital Culture archive, allowing the inclusion of less subject-driven materials such as university archives in the geographic South, in addition to the more traditional subjects of this region, including the Civil Rights Movement, the railroad industry, slavery, and the Civil War.

Decisions around the comprehensiveness of scope may be achieved through other means as well. The content contributors may choose to collect materials relating to a specific topic, but restrict it by time period, geographic region, or genre. It may be important to reserve leeway for contributing institutions to include materials that might not strictly fit the criteria of the network but, nevertheless, warrant preservation. The founding principles of a specific DDP network can determine how narrowly or broadly to define the scope of the archive.

The collections of the contributing institutions need not be homogeneous to create a successful network; however, by establishing a common scope, a DDP network may develop a common sense of purpose among the content contributors that are jointly investing in preserving each other's collections.

After determining the scope of the preservation archive, the next step is to document what content is on-target, and how fluid the definition is, and if necessary, to consider circumstances for expanding the scope. These early decisions made and documented by the members of a DDP network may not only affect the rate at which the network grows, but also its ability to attract new members.

Copyright Status

Copyright, which will be discussed further in Chapter 8: Copyright Practice in the DDP, must be considered when establishing a DDP network and selecting digital materials to preserve.

Many preservation efforts conflate maximizing short-term access (i.e., high availability) with long-term access (i.e., preservation). High availability entails adopting strategies for ensuring that content is constantly available to the public. It also mandates that content is free of copyright and intellectual property constraints through the use of appropriate licenses or permissions owned by the contributing institution.

A DDP network may be an open archive, or it may reside somewhere on the spectrum from dim to dark archive. That is, it may be open to only the contributors' servers for ingesting (dark archive); it may be open to specified users, such as the contributing institutions' communities (dim archive); or it may provide unrestricted access (open archive). This status will determine whether contributors will focus solely on long-term preservation issues, or some combination of preservation and public access issues.

- **Open PLN Archives:** CLOCKSS (Controlled LOCKSS) is a not-for-profit, community-governed, alliance of research libraries and publishers.[6] Though somewhat different from many of the PLNs explored throughout this book due to its journal content targeted for preservation, CLOCKSS is an excellent example of an open archive. For example, when a trigger event has occurred and the digital content is no longer available from a publisher, one of the participating institutions will move the content to a hosting platform and the impacted preserved content will be made available without charge to the world.

- **Dark PLN Archives:** Neither the MetaArchive Cooperative[7] nor the Alabama Digital Preservation Network (ADPNet)[8], as dark archives, has a public access component at the present time. Preservation and access, though united in their goals, are considered two separate functions. Only content contributors in the PLN have access to the collections, and this access is restricted to ingesting collections into preservation caches and to restoring digital content to the contributor (i.e., not to view or use the content). The contributing institution determines whether access is provided or not.

- **Dim PLN Archives:** The original LOCKSS public network provides preservation and access to content governed by the legal or license agreement associated with that content. For example if a subscription publisher limits access to a range of IP addresses, access to that publisher's LOCKSS preserved content is limited to the same range of IP addresses. Government documents are also preserved in the LOCKSS system. This content is not subject to any further access restrictions either from the publisher or from the LOCKSS system.

Whether or not the preservation network accommodates public access to the preserved content, each member institution must be responsible for implementing appropriate standards for addressing copyright, intellectual property, and issues related to content that has been contributed. Content contributors bear the responsibility for determining ownership and their rights to preserve the content prior to submitting it to a DDP network. Compliance with laws – including the use of exemptions set forth within U.S. Code Title 17 (copyright law) in sections 107, 108[10], and elsewhere, and permissions through deeds of gift or other clearances is an obligation of each institution in the PLN. International institutions must similarly address intellectual property and copyright laws. Rights should be documented in the collection-level metadata.

Preservation networks rely on a great deal of trust, including the trust that contributing institutions are not violating copyright law when sharing their digital files, even for preservation purposes. Trust needs to be formalized in a legal agreement indicating that contributors represent and warrant that, to the best of their knowledge, they are not contributing content to the preservation network that would infringe the rights of others. Each contributor should also certify that it holds sufficient rights to authorize the DDP network to use the content in a manner consistent with the requirements of a multi-cache preservation strategy, whether it is a dark archive or one that provides some level of public access.

The Membership Agreement for the MetaArchive Cooperative is an example that covers these formal issues with appropriate terminology and legal language. [11]

Risk Factors

One of the major concerns when pursuing digital initiatives is the fear of loss due to many potential factors, including natural disasters, human errors, fires, floods, power surges, and more. However, previous worries about unstable media and hardware obsolescence have been greatly reduced after more than two decades of providing digital media to library constituencies. An excellent grounding in these issues is available in the Council on Library and Information Resources' 2000 publication, *Risk Management of Digital Information: A File Format Investigation*.[12] It outlines a variety of factors that might put a digital collection at risk, and supplies a pragmatic approach to assessing risks of digital collections with its "Risk-Assessment Workbook." The DRAMBORA assessment tools likewise supply institutions with a workbook approach to risk assessment and management.[13]

In addition to the safe harbors that are created in a DDP network, it is also important that the contributing institutions in the network make decisions based on long-term access goals [or strategies], not just current technology. Once the content contributors have agreed on the risk factors for their collections, they can assign priorities for ingestion into the network using risk rankings. Because not all content can be ingested simultaneously, and not all content may be worth preserving, each DDP network may wish to set risk guidelines to prioritize content for ingest. They might also review where files are stored, and on what type of media. For example, large digital master files may exist solely on external media, such as compact discs (CDs). In order to be ingested by the other members of a PLN, the files would need to be transferred onto a web server and arranged into archival units (AUs). Files stored on servers are likely to be safer in most cases than those on offline storage media such as CDs and DVDs. Finally, it might consider whether a file has been backed up, and if so, whether those backups are tested regularly.

With these issues in mind, consider the risk levels adopted in 2004 by the MetaArchive Cooperative as it launched its Southern Digital Culture archive. In 2009, the Cooperative still uses these guidelines for this archive, and has extended them so that they may also be applied to new archives established by the Cooperative:

1. **Extreme Risk:** No one is responsible for preservation. No other copies of the digital content are preserved. No regular backups or data migration.
2. **Significant Risk:** Responsibility under discussion, but no copies of the digital content are currently being preserved.
3. **High Risk:** Only one backup of digital masters on CD-ROM. No regular backups or data migration.
4. **Moderate Risk:** Some danger that collection backups might be lost in the future.
5. **Low Risk:** Copies are backed up regularly with a long-term maintenance plan in some other trusted digital archive.

CONTENT INGEST PREPARATION

Organizations create digital collections as part of their ongoing work, but often ignore or set aside long-term planning, which results in idiosyncratic and ad-hoc data storage structures. Such early idiosyncrasies can become embedded in these collections' data structures, upon which digital infrastructure and management workflows continue to be built. Such infrastructures may cause prodigious problems during systematic efforts to preserve the content of these (static or growing) collections.

This section outlines two important components, one that is broadly applicable to DDP networks and one that is specific to PLNs. First, it will outline how to prepare a collection to be programmatically ingested into a DDP network, and then it will specify how to initiate content ingest within a PLN. It provides examples of up-front planning with long-term preservation in mind, including clearly defined and documented collection data structures. It also suggests remedies for collections that evolved with little or no direction.

Content ingest requires the following elements:

- Accessibility (for PLNs, this must occur using the Web)
- Organizing Collections
    - Data Wrangling
    - Metadata Creation
- Defining Archival Units for a PLN Solution
- Manifest Page Creation for a PLN Solution
- Plugin Creation for a PLN Solution

Each of these elements is described below in greater detail. For more information about basic DDP and PLN architecture, please see Chapter 2: DDP Architecture. For additional details on preparing content for ingest into a PLN network, please refer to Chapters 6: Content Ingest, Monitoring, and Recovery.

Accessibility

For any DDP network to ingest/harvest content, it must first be made accessible to that network. This may occur through a submission process in which the contributing institution sends files to a central location or it may happen through web-based harvesting or other mechanisms.

For example, in order for content to be ingested into a PLN, it needs to be web-accessible via HTTP (Hypertext Transfer Protocol) or HTTPS (secure HTTP). When access restrictions are in place (e.g., only constituents from the contributing institution have access), a list of specific preservation members' IP addresses must be added to the web server's firewall configuration to enable ingest by the authorized PLN institutions. For more details on this configuration please see Chapter 7: Cache and Network Administration for PLNs

Organizing Collections

For preservation and life-cycle management purposes, a digital collection and its content should be clearly arranged, defined, and described. When beginning a digital initiative, it is wise to consider what might be necessary for both programmatic capture and online user access, such as hierarchical arrangement and logical file naming (see the section on "Content Management" below).

When creating a new digital collection, it is highly recommended that an institution organize it into a methodical or hierarchical file structure. For example, an Electronic Theses and Dissertations (ETD) collection may require a new directory for each submission year. Digitized special collections could follow the same organizational structure as the physical collection, which often has a hierarchy of folders within a series. Naming conventions should include logical labels for each folder in each series. The series can be organized by subject, as well as chronologically or alphabetically. Even when only a portion of a collection is digitized, a complementary file directory structure should be established to better manage the long-term preservation of the digital items. This practice will avoid the creation of a directory that is a hodgepodge of files. Other logical arrangements could resemble a business organizational structure, a genealogical family tree, or a calendar of events. Documenting any policy that is developed helps to ensure its understanding and usage by future digital collection managers.

For the purposes of the MetaArchive Cooperative's PLN, a collection is defined as the aggregated content to be preserved under the banner of one collection-level metadata record, which is entered in the Cooperative's conspectus database. (see the section on "Metadata," below.) It may differ from the original analog or digitized collection because the entire collection may not be digitized or digitally preserved due to copyright, risk, or other reasons.

Data Wrangling

It is not atypical to encounter existing digital collections that were created without forethought, resulting in rather haphazard collections that are not preservation-friendly. Data wranglers alleviate these problems by wrestling the digital objects into discernable units.

When associated with a PLN, data wrangling refers to the strategic rearrangement of digital collections so that the path to them can be logically defined for programmatic access. In order for the content to be ingested into the PLN (which uses web-based mechanisms for this ingest process), some data wrangling may be required to assemble the files into a coherent order (or to identify their location) and to describe the collection clearly and thoroughly for effective future access. This effort has been particularly necessary for older collections established in the early days of the Internet, when making them electronically accessible was often rushed and not approached in a strategic, long-term manner.

Data wrangling may entail moving and rearranging master files and metadata into directories and folders corresponding to newly created file directories for the collection and its sub-collections (or, in the PLN context, its Archival Units (or AUs, as described in "Defining Archival Units" below) This inevitably leads to discoveries of missing, mis-numbered, duplicated, substandard, or corrupted files, as well as insufficient metadata. Identifying and correcting these errors will aid not only in preserving the digital assets, but also in providing both short- and long-term access to them.

Qualified staff members who know the custodial history (provenance) of the materials to be preserved should make the decisions about arrangement and description. However, university members of PLNs have found student employees to be effective data wranglers, preparing collections for ingest by moving files or creating virtual collections. As described further in Chapter 4: Organizational Considerations, data wranglers may also write plugins and manifest pages to permit digital content to be ingested by LOCKSS. This preservation work is sometimes analogous to processing physical archival collections, which must be arranged, inventoried, and re-housed before they can be accessed.

Metadata

Digital preservation depends in large part on ascribing effective metadata (structural, technical, and descriptive) to objects and collections. DDP networks, including PLNs, have to make choices about what metadata standards they wish to employ, and at what level: the network or contributing institution. That metadata aids preservation is an uncontested principle; however, metadata standards can become a barrier to entry for potential network participants. Each DDP solution must weigh the pros and cons of such metadata standards as PREMIS and METS, and must determine what level of standard best suits the preservation needs of its member institutions.

For example, the MetaArchive Cooperative has found that collection-level metadata is an essential tool for its preservation network, as it facilitates tracking and maintenance of the content. Contributors with backgrounds in archives, systems, cataloging, and digital libraries can be helpful in fully describing collections in ways that are meaningful to both the contributing institution and to the network monitoring process. It is important that they not only have knowledge of the collection, but also understand the preservation goals and functionality of the PLN. Detailed information about each ingested collection also facilitates network management and assists with various access-related issues, including disaster recovery, where a contributor needs to use the preserved digital content to rebuild its local collection. The Cooperative does not, however, require its contributing institutions to limit their preservation activities to those collections that have item-level metadata in any particular schema, as the differing practices of its member institutions means that any such requirement would necessarily limit the preservation of their collections.

There are a number of excellent existing schemas that can be used or adapted to meet a DDP network's collection-level metadata needs, including the following:

- BCR CDP's Dublin Core Metadata Best Practices[14]
- Dublin Core Collections Application Profile[15]
- UKOLN Research Support Libraries Programme (RSLP) Collection Description Schema[16]
- IMLS DCC Collection Description Metadata Schema[17]
- PREMIS Preservation Metadata: Implementation Strategies[18]
- MetaArchive Collection-Level Conspectus Metadata Specification[19]

Each schema contains standard elements for library and archival description, such as title or creator. Some metadata elements in these schemas are tailored specifically to digital objects, such as MIME format. As a DDP network considers which elements to include in its schema, it should think about how it wants to record, and in what order the materials will be ingested, as well as information regarding accrual, or how often a particular collection is updated. These elements are important for ingest and for storage projections. Depending on the needs of the DDP network, the collection description can require controlled vocabulary (e.g., Library of Congress Subject Headings) or code (e.g., ISO 639-2 language code).

Continuing with the MetaArchive Cooperative's example, the Cooperative determined that there are eight principal categories of metadata elements:

1. Descriptive data illustrates or explains the collection.

2. Uniform resource identifiers (URIs), uniform resource names (URNs), and unique identifiers locate the collection.

3. Coverage places the collection in space and time.

4. Accrual information anticipates the growth of the collection.

5. Data description provides formats, sizes, languages, etc.

6. Rights and ownership elements document intellectual property and provenance.

7. Related resources inform about associated collections.

8. Ingesting information provides data necessary for the ingest process.

In order to identify, ingest, and track the collections of a DDP network, each contributor may record collection-level administrative and descriptive metadata in a DDP-specific database (in the case of the MetaArchive Cooperative this is the conspectus database, which is freely available to other PLNs). This database describes the breadth of the DDP network through network-wide and institution-level views. Each collection, which in the PLN arena may be comprised of one or more AUs, has one corresponding metadata record in the database. The database should provide metadata versioning support to track collection changes.

The conspectus database designed by the Cooperative interoperates with the LOCKSS title database, providing relevant information in an XML dialect of RDF. The title database contains the XML parameters that tell the LOCKSS daemon three central things: 1) where to find plugins as signed jar files, 2) the location of archival units, and 3) the list of IP addresses for caches participating in a network. The consistent use of XML makes it easier for the conspectus database to generate the title database as well. To this end, the conspectus also records metadata that is required for ingest by the LOCKSS software: the plugin name, plugin parameters (where used), and the base URL of the collection.

The Cooperative also recommends preserving local item level descriptive, administrative, and structural metadata for the digital objects in the collections wherever such metadata exists. The metadata should be in a sustainable format such as unformatted (ASCII) text or XML, and should be ingested by the PLN along with the collections they describe.

Defining Archival Units for a PLN Solution

As described in more detail in Chapter 6: Content Ingest, Monitoring, and Recovery, PLN ingests are conducted through guided crawling, which is much more exact than typical web-spidering methods. PLN ingests target specific collection components based on their Archival Units (AUs) — which are the collection boundaries established by the content curator before a given collection is slated for ingest.

AUs are the building blocks of a LOCKSS collection. An AU is a cohesive and logical aggregation of content by topic, format, file size, or file location that is intended to divide a collection into discrete groupings (typically between 1 GB and 20 GB in size) for ingest into the PLN. For example, each AU of a collection of digitized yearbooks might comprise a single volume, or a collection of ETDs could have AUs for each year's theses and dissertations.

AUs for large digitized manuscript collections may correspond to the hierarchical folder arrangement. That is, the files and metadata may be organized into record groups, boxes, folders, sub-folders, and items. For a collection of photographs, an AU might be the entire collection, or it might be a folder containing the digital masters for the collection. If size and organization permit, an AU can encompass all items in an entire record group.

Examples of AUs include:

- One volume of an e-journal
- One year of ETDs
- One decade of scanned yearbooks
- One folder of archival TIFF images or sound files

Manifest Pages for a PLN Solution

Each AU must have a manifest page, which serves as a starting point for ingest, and a statement granting permission to LOCKSS to ingest the AU. The manifest page is usually a normal HTML page, and must link (usually indirectly) to all the content that should be included in the AU.

The permission statement is usually contained on the manifest page, but it may be located anywhere on the same host as the content to be ingested. Either the following statement: "LOCKSS system has permission to collect, preserve, and serve this Archival Unit", or a Creative Commons license, is acceptable. The statement need not be visible to users, e.g., it can be placed within an HTML comment.

Best practices for manifest pages include:

- Making sure AUs are properly accounted with an individual manifest page, or a collection-level manifest page.
- A manifest page should avoid, when at all possible, attempting to encapsulate a complete list of files that are to be ingested. It should instead point to the location of the AUs.
- Although not required by LOCKSS, it will assist long-term preservation efforts if each manifest page contains the name of the collection, the institution, and a contact name/address, and is updated to reflect changes in this administrative information as well as changes in the AUs.
- Manifest pages should contain a short description of the structure of the collection, such as where to find

metadata, the naming conventions used in filenames, and how the AUs relate to the site structure.

- A collection's manifest page should contain a link to its collection-level metadata description.

Plugins for a PLN Solution

A plugin provides information to LOCKSS about a collection or a group of similarly structured collections to tell it how to collect and audit the content. A plugin is a small block of XML which, when given a set of AU-specific parameter values (such as base URL and year), defines the URL of the AU's manifest page.

LOCKSS has a plugin tool[20] and a plugin tool tutorial[21] freely available online. In addition, the MetaArchive Cooperative has created a Plugin Standards Checklist to guide the plugin creator through the Java coding decisions.[22] Virginia Tech has also produced a plugin tutorial that contains a case study on ETDs.[23]

As plugins are under development for a collection, they should be stored in a separate plugin repository or repositories. Plugin repositories can be housed and managed centrally for the entire PLN, locally, or a combination of the two. The MetaArchive Cooperative, for example, has deployed its plugin repository in a cloud computing environment, which allows for centralized location of the plugins, but provides a decentralized shared location for access and submission. The plugin repository or repositories should be placed under some kind of version control. LOCKSS makes use of a Concurrent Versions System (CVS), and the MetaArchive Cooperative makes use of Subversion. This allows for on-going changes to be documented, and if necessary to revert to earlier configurations.

Plugins must also be tested on a test cache or a test network to ensure that their crawl configuration successfully ingests the collection. This test network will be discussed in greater detail in Chapter 6: Content Ingest, Monitoring, and Recovery.

Once created and tested, plugins are packaged into files called JARs (an acronym for Java™ ARchive), which are signed and stored in a final plugin repository that is accessed by LOCKSS before it initiates an ingest on a cache. The LOCKSS daemon initiates ingest by accessing the title database to locate the proper plugin, its parameters, and the base URL from which to begin collecting or re-crawling content.

CONTENT MANAGEMENT

A Case Study in Preparing Content for LOCKSS Preservation

In the process of accumulating digital collections it is normal for directory structures, naming conventions, and metadata forms to become highly idiosyncratic, outmoded, and a hindrance to preservation readiness. When the focus turns to digital preservation readiness, then institutions become aware of the long-term detrimental effects of ad hoc preparedness.

For example, in 2008, the MetaArchive Cooperative and the Networked Digital Library of Theses and Dissertations (NDLTD) formed an alliance to examine the practical issues involved in a collaborative replication strategy for the digital preservation of ETDs. Shared below, the findings from that effort help to clarify the need for digital preservation readiness.

Preserving Restricted and Withheld ETDs

As previously mentioned in the section titled "Accessibility" above, to add a digital object to a PLN, it must be web-accessible (i.e., available via the HTTP or HTTPS protocol). PLN ingest requires a standard HTML permission-to-preserve statement on the host containing the ETD directory on its manifest page (see section titled "Manifest Pages" above).[24] While the manifest page is human readable, it is used entirely for programmatic ingest by the preservation network contributors. When access restrictions have been placed on some ETDs (for example, host university-only access), a list of specific content contributors' IP addresses can be added to the web server's firewall configuration to allow ingesting by only the specific network caches.

Structuring New ETD Collections for Ingest and Recovery

Organizing a contributing institution's ETDs most effectively for preservation ingesting relies upon the creation of a methodical structure, such as a directory for each year's ETDs. For larger institutions that approve hundreds of ETDs each year, annual directories should be further subdivided into logical units such as semesters or months. Smaller institutions that approve 100 or fewer ETDs per year will not benefit from creating these subdirectories.

While structures optimized for human browsing might be based on departments, authors, advisors, etc., an organizational approach designed for comprehensible workflow and preservation of a growing collection is more usefully based on accumulation periodicity. Adopt a common, easy-to-decipher naming convention; for example, year/month 2008/01, 2008/02, etc. Remember, however, that these units are for programmatic ingest and not for human browsing.

When every contributing ETD member follows the same conventions for directory structure and file naming, each collection can be handled by a single plugin with different base URL and year parameters (see the section titled "Plugins" above). This consistency enables the network to provide members with effective but generic plugins. Otherwise, each institution must generate a plugin specific to its structure. The goal is thus to standardize naming conventions for files and directory structures from the beginning of any project. This will require analyzing the ways that the collection may grow over time, scoping numbering systems that can be parsed automatically, and developing of directory structures that can be easily traversed by subsequent ingesting systems. Data structures should also ideally be aligned with item-level metadata (see the section title "Metadata" above).

Successful ingest will depend on the content contributor's ability to structure content into manageable AUs (see the section titled "Defining Archival Units" above). Each contributing institution needs to consider how the preservation copies stored in the network might be used to repopulate its existing archival structure in the future. A benefit of LOCKSS requires institutions to address preservation readiness at the point of ingest. For the network to easily ingest the content, a contributing institution is advised to have that content organized and well structured. Contributing institutions should remember that whatever work they have done to export the files and folders out of its repository system will need to be done in reverse in order to use them to repopulate that system.

URNs for ETDs

As students submit their ETDs, the files should be assigned a unique directory identifier with a Uniform Resource Name (URN). For example, an ETD submission that began at 5:57:13 on March 7, 2001 might become etd-03072001-175713/, based upon the date and time of submission. In this example, the ETD submission began at 5:57:13 on March 7, 2001. After an ETD is approved, the file(s) become part of the local collection. If an ETD that requires temporary embargo is approved, the upper directory structure would be somewhat different, but the URN would be structured in the same predictable way. For example, an effective plugin would direct ingest from a given URL to find all the 2001 ETDs with instructions to:

1. Ignore the four numerals and '-' immediately following "etd" (i.e., -0307),
2. Recognize the year (in this case, 2001); and
3. Ignore the remaining characters.

This ETD, whether it has one file or many, would be placed in the school's 2001 AU. With this process, each year's-worth of ETDs is readily identifiable from each URN and can be divided into AUs by year on the preservation network caches without any data wrangling.

Triage for Legacy Collections

But what about collections that have been subjected to multiple repository conventions and those that straddle the gap between digitized and born-digital ETDs? Using Virginia Tech (VT) as a case study, the following demonstrates remediation approaches for entrenched ETD collections.

ETDs approved at VT before 2000 were named using a variety of URN conventions, such as /etd-454016449701231/ and /etd-030999-145545/. These URNs were not clearly structured or consistent though etd-030999-145545/ was probably submitted in 1999 and etd-454016449701231/ was most likely submitted in 1997. The solution was to establish a virtual collection with one AU for all pre-2000 ETDs. Plugin instructions were set to find all ETDs that did not fit the post-1999 URN convention. The complexity of this largely static collection is ultimately best served by plugin rules that exclude anything that matches the post-1999 format and places it into the PLN in an "Early VT ETD Collection."

Because digitized (as opposed to born-digital) bound theses and dissertations (BTDs) often follow the establishment of an ETD initiative, there exists a welcome opportunity to learn from earlier experiences. Scanned theses and dissertations can follow the URN naming convention based upon their digitization dates, rather than the dates on which they were originally approved. For example, a dissertation that was completed in 1994, but scanned Oct. 2, 2007 at 2:48:46 would be ingested with the existing plugin and preserved in an assigned AU with the born-digital ETDs submitted in 2007. This method allows the static collection to remain unchanged. This system works for preservation purposes; however, it may need further consideration for rebuilding a public ETD database or collection from the preservation cache because works will likely be difficult to programmatically identify when reestablishing an annual grouping based on year of completion/approval.

It would not be very complicated on a conceptual level to programmatically generate URNs for BTDs based on their completion date, as this information likely exists in the contributing institution's MARC bibliographic records. BTDs are often assigned Library of Congress call numbers that also include dates. For example, the Limoges dissertation has the call number LD5655.V856 1994.L556. These call numbers are constructed as follows: Institution number--LD5655, thesis/dissertation number--V856, year--1994, Cutter number--L556.

In addition to file naming, batch processing involves pulling the physical items from possibly multiple locations (e.g., main library and remote storage). The process of arranging and maintaining their order, accurately deriving the file names from the MARC records, and linking them to the appropriate BTD files would become overly cumbersome and inefficient.

Final Remarks

Some PLNs, like the MetaArchive Cooperative, separate the function of the preservation caches from locally accessible collections. If it becomes necessary to rebuild a database of digital theses and dissertations (both digitized and born-digital) at the originating institution, the restoration of access, arrangement, and/or display of ETDs and BTDs is largely external to the purpose of the PLN. The goal of this case study has been to highlight the importance of organizing digital collections in ways that optimize both ingesting and repopulating a contributor's collections in the event of catastrophic loss – acknowledging that in every extant preservation solution there are going to be some trade-offs. The benefit of the LOCKSS solution is that it not only requires a contributor to become proactive in their digital preservation readiness, but also provides them with a sufficient amount of flexibility to carry out the preservation in ways that are best suited to their content and priorities.

CONCLUSION

In order to ensuring the successful ingest of content into a PLN, a content contributor must pay careful attention to its content's structure prior to its submission for harvest. This chapter has stressed the importance of pursuing preservation readiness before pursuing preservation itself. In the next chapter, the benefits of this expended effort will become more apparent as the process of collection ingest, monitoring, and recovery is covered in greater detail.

ENDNOTES

1. Library and Archives Canada's Digital Repository Services and Standards Office, *The Unified Digital Formats Registry*, 2009 http://www.udfr.org/ (last accessed 12-14-2009).
2. Library of Congress, "Sustainability of Digital Formats: Planning for Library of Congress Collections," *Digital Preservation*, 2007 http://www.digitalpreservation.gov/formats/ (last accessed 12-14-2009).
3. Rog, Judith; van Wijk, Caroline, "Evaluating File Formats for Long-term Preservation," *Koninklijke Bibliotheek*, 2008, 2. http://www.kb.nl/hrd/dd/dd_links_en_publicaties/publicaties/KB_file_format_evaluation_method_27022008.pdf (last accessed 12-14-2009).
4. Digital Library Foundation/OCLC Registry of Digital Masters Working Group, "Registry of Digital Masters Record Creation Guidelines," *The Digital Library Federation*, 2007 http://www.diglib.org/collections/reg/DigRegGuide200705.htm (last accessed 12-14-2009).
5. MetaArchive Cooperative; Networked Digital Library of Theses and Dissertations, "Electronic Theses and Dissertations Preservation Survey," *MetaArchive Cooperative*, 2008 http://www.metaarchive.org/public/resources/ndiipp_docs/NDIIPP_Market_Analysis.pdf (last accessed 12-14-2009).
6. CLOCKSS (Controlled LOCKSS): http://www.clockss.org/clockss/Home (last accessed 12-14-2009).
7. MetaArchive Cooperative: http://www.metaarchive.org (last accessed 12-14-2009).
8. Alabama Digital Preservation Network (ADPNet): http://www.adpn.org/ (last accessed 12-14-2009).
9. LOCKSS (Lots of Copies Keep Stuff Safe): http://www.lockss.org/ (last accessed 12-14-2009).

10. The Section 108 Study Group has been charged with updating for the digital world the Copyright Act's balance between the rights of creators and copyright owners and the needs of libraries and archives. This site has both an Executive Summary and the Full Report of the Study Group: http://www.section108.gov/ (last accessed 12-14-2009).

11. The Membership Agreement of the MetaArchive Cooperative is a formal legal agreement for members to represent and warrant that, to the best of their knowledge, they are not contributing content to the preservation network that would infringe the rights of others: http://www.metaarchive.org/sites/default/files/Membership_Agreement_2010.pdf (last accessed 12-14-2009).

12. Lawrence, Gregory W.; et. al., "Risk Management of Digital Information: A File Format Investigation," *Council on Library and Information Resources*, 2000 http://www.clir.org/pubs/reports/pub93/contents.html (last accessed 12-14-2009).

13. Digital Repository Audit Method Based on Risk Assessment (DRAMBORA): http://www.repositoryaudit.eu/ (last accessed 12-14-2009).

14. The intent of the CDP Dublin Core Metadata Best Practices (CDPDCMBP) is to provide guidelines for creating metadata records for digitized cultural heritage resources that are either born digital or have been reformatted from an existing physical resource. This document uses the Dublin Core element set as defined by the Dublin Core Metadata Initiative (DCMI): http://www.bcr.org/dps/cdp/best/dublin-core-bp.pdf (last accessed 12-14-2009).

15. The Dublin Core Collections Application Profile specifies how to construct a collection level description. It provides a means of creating simple descriptions of collections suitable for a broad range of collections, as well as simple descriptions of catalogues and indexes. Aggregations of physical or digital resources (collections) and aggregations of the metadata that describe them (catalogues and indices) can be described with similar properties: http://dublincore.org/groups/collections/collection-application-profile/2007-03-09/ (last accessed 12-14-2009).

16. UKOLN Research Support Libraries Programme (RSLP) Collection Description Schema proposes a collection description schema, a structured set of metadata attributes, for describing collections within the RSLP: http://www.ukoln.ac.uk/metadata/rslp/schema/ (last accessed 12-14-2009).

17. The IMLS DCC Collection Description Metadata Schema is based on the UKOLN RSLP Collection Description Metadata Schema and the Dublin Core Collection Description Application Profile. The IMLS DCC project has adapted these schemas to reflect the particular

nature and needs of the included projects. It is meant to describe digital collections created through IMLS Grant projects and does not describe in detail the projects themselves. This metadata schema forms the basis of the IMLS DCC Collection Registry: http://imlsdcc.grainger.uiuc.edu/CDschema_elements.asp (last accessed 12-14-2009).

18. PREMIS Preservation Metadata: Implementation Strategies is an initiative aimed at defining a core set of semantic units that repositories should know in order to perform their preservation functions: http://www.loc.gov/standards/premis/ (last accessed 12-14-2009).

19. The MetaArchive Collection-Level Conspectus Metadata Specification charts and defines in detail the thirty-five elements that describe each collection: http://metaarchive.org/sites/default/files/conspectus_md_2005.html (last accessed 12-14-2009).

20. The following is a link to the LOCKSS plugin generation tool, which provides a user interface for creating and testing a plugin: http://www.lockss.org/lockss/Plugin_Tool (last accessed 12-14-2009).

21. The following is a link to brief overview of how to use the LOCKSS Plugin Tool: http://www.lockss.org/lockss/Plugin_Tool_Tutorial (last accessed 12-14-2009).

22. http://metaarchive.org/metawiki/index.php?title=PluginStandards (last accessed 12-14-2009).

23. The following is a link to a mini LOCKSS Plugin Tutorial that was developed by Virginia Tech graduate student Kamini Santhanagopalan: http://scholar.lib.vt.edu/lockss/introduction.htm (last accessed 12-14-2009).

24. For an example of a manifest page in the context of the discussion on "Preparing Content for LOCKSS Preservation" see this link to the Virginia Tech ETDs LOCKSS Manifest Page: http://scholar.lib.vt.edu/theses/lockss/manifest.html (last accessed 12-14-2009).

Chapter 6: Content Ingest, Monitoring, and Recovery for PLNs

Katherine Skinner (Educopia Institute)

Matt Schultz (Educopia Institute)

Monika Mevenkamp (Educopia Institute)

OVERVIEW

This chapter details how content is ingested, monitored, and restored in the PLN context. These technical steps are critical for processing and preserving the content that institutions submit to a PLN. They are the shared responsibility of the content contributor and the other members of the preservation network that ingest and maintain copies of these collections.

These steps are governed either by the PLN staff (e.g., the MetaArchive Cooperative) or by the LOCKSS staff (e.g., ADPNet). The tools that networks use to complete these tasks may differ across different PLN setups. This chapter focuses primarily on the procedures and tools used by the MetaArchive Cooperative in a self-administered network context. All tools described below are freely available open source components that are either standard to the LOCKSS infrastructure or have been built as modules that work in conjunction with the LOCKSS system.

This chapter draws heavily upon the work of the MetaArchive Cooperative and includes many recommendations based on our five years of experience running a PLN.

CONTENT INGEST

This section details the final stages of preparation that occur before content is ingested into the PLN, namely proper plugin development and testing, the use of test caches or networks, and completing title database entries for a collection via a conspectus tool. It then describes the processes of designating what caches should ingest the content, alerting those caches that content is ready for ingest, and ingesting the content at the local cache level.

Plugin Development and Testing

Plugin parameters must be defined accurately in order to guide the LOCKSS daemons to ingest only and all of the content that a content contributor intends to preserve. If a plugin contains errors, the daemons will either not be able to crawl the content or, worse still, will crawl and ingest the wrong content. Such scenarios are far from ideal, particularly since LOCKSS (intentionally) does not provide a central means of deleting content from the network. As a result, testing plugins is a central step in the content contributor's ingest workflow.

Initially, the MetaArchive Cooperative encouraged developers at contributing institutions to develop and test their plugins using their own tools and procedures. It maintained multiple plugin repositories and allowed content contributors to manage the testing process largely for themselves. This approach yielded a mixed result; some content contributors produced solid plugins, and others introduced erroneous plugins into the network.

As it worked to streamline its operations over the last two years, the Cooperative decided to standardize the process of writing and testing plugins as much as it could through providing plugin templates, developing documentation, and making use of a central plugin repository with discrete development, testing and release branches. This move has simplified plugin creation at both the content contributor and the network levels. The content contributor is still responsible for writing and testing its plugins, but it now has a centralized infrastructure in which to do so.

Based on this experience, the Cooperative recommends that other independent PLNs consider implementing one plugin repository that is placed in a centrally accessible location under some form of version control to facilitate the development and testing process. It also recommends that networks create standard testing procedures, preferably in a test network as described below.

Test Network Best Practices

A plugin should enter the network only after a content contributor has tested that plugin and determined that it produces the expected crawl results. The test network can be implemented locally or centrally. In the local scenario, each content contributor uses a temporary instance of the title database and runs a manual command on the LOCKSS daemon in order to test a plugin. The network can alternately implement the test network as a central resource for all members to use. The content contributor can then verify the accuracy of a plugin through the test network's cache manager or the LOCKSS daemon user interface on their test cache.

Re-Testing Changed Collections

Whenever a content contributor makes structural changes to content post-ingest (including changes to the content storage structure), it must ensure that the new or altered content can be ingested properly into the network. The modified content must either fall within the previously defined collection parameters or the collection parameters will need to be redefined in the plugin to encompass the new or modified material.

If a collection directory structure or location changes, the content contributor may need to create new archival units and/or revise the collection's plugin to provide the proper parameters and location to aid the daemons as they continue re-ingesting the content at regular intervals.

Any changes made to any plugin for any reason necessitates a new round of testing on a test cache or in a test network. For best practices on how to mitigate against the need to edit a plugin, please refer to the section on Content Management in Chapter 5: Content Selection, Preparation and Management.

Finalizing the Title Database

Once a successful test ingest has been verified, the content contributor must signal to the rest of the PLN that the content is prepared and readied for ingest into a production network. Independent PLNs may use the conspectus tool developed by the MetaArchive Cooperative to facilitate this process.

The Conspectus Tool

The conspectus tool is a web-based data management tool that maintains both LOCKSS-specific technical metadata and descriptive collection-level metadata about each digital collection that is submitted for ingest. Content contributors can use the conspectus to create, update, and maintain their collection descriptions. The conspectus also generates LOCKSS-specific configuration data that is stored in the title database and is used by system scripts to configure the network. The conspectus contains a metadata schema designed by the Cooperative that includes elements from widely used metadata standards such as Dublin Core, METS, and MODs.

The content contributor prepares a collection description in the conspectus for each collection, providing a name and title for the collection and entering collection-level metadata. The content contributor also enters collection configuration parameters (information about the archival units and the plugin). The content contributor reviews the collection, its metadata, and its ingest information to ensure that the content and its context are properly defined for preservation purposes.

Designating Caches for Ingest

Once the content contributor designates the collection as ready for ingest, it publicizes the crawl procedure for the collection by moving the tested and approved version of their plugin to a release branch of the plugin repository (as described in Chapter 5). Because PLNs are configured with a distributed infrastructure in which each cache is autonomous (i.e., even any centralized staff would not have direct access to every cache), there is no automated way to initiate a crawl of content by every cache in the network. Each PLN must, therefore, establish a central way to alert each cache when a new collection is ready to be ingested. If the network is large enough that not every preservation cache needs to ingest the content, the PLN also needs to determine how to make this alert specific for only those caches designated to harvest each collection.

Communication Best Practices

Project listservs and conference calls can provide an effective alert strategy for designating sites for replication and signaling the ingest process. For example, in the MetaArchive Cooperative context, the assignment of caches and the signal to ingest is accomplished through a combination of both mechanisms using the following sequence of events:

1. The content contributor notifies the PLN's central systems administrator that the collection is ready to be ingested.

2. The central systems administrator verifies that the content contributor has completed a successful test ingest and a title database conspectus entry, and requests that the PLN program manager designate seven sites to ingest the content. These sites are typically chosen on the basis of cache capacity, fairness regarding the amount an institution

contributes vs. the amount it hosts, and geographical distribution.

3. The program manager verifies these locations with the central system administrator.

4. The program manager then formally designates these seven sites using the conspectus database. The program manager also contacts all designated caches via a program listserv to alert them that there is a new collection to crawl.

5. Each designated member's cache administrator initiates a crawl by adding the identified collection to its cache's configuration through their LOCKSS daemon user interface, and monitors the crawl to ensure its completion.

6. The PLN's central system administrator, the content contributor, and the designated cache location then take joint responsibility for ensuring that the collection is properly crawled and ingested successfully by each designated cache (using the cache manager), and all parties report back on weekly conference calls to confirm a successful ingest or to discuss any issues that may have arisen on those caches.

As soon as the content is ingested into the network, the LOCKSS software begins actively evaluating and preserving that content as discussed in greater detail below.

CONTENT PRESERVATION

This section covers the process by which caches meet agreement on the completeness and correctness of content, the difference between active and closed collections, and the importance of configuring plugins and web servers for effective network preservation.

Polling

When all designated caches have completed an ingest process, these caches regularly engage in polls in which they compare the ingested files (via cryptographic hashes) to ensure that all copies of the content match.

In a successful polling environment, if the content in one cache does not match the others, the LOCKSS daemon identifies the inconsistent cache and triggers a re-crawl of the content from what is considered the authoritative copy—the content contributor's site. If the content contributor site is no longer available (as described below), the inconsistent cache's content is repaired by one of the other preservation caches.

Active vs. Closed Collections

Content contributors have two options for ongoing preservation activities for each collection they submit. They can either request that the LOCKSS daemons actively revisit the authoritative copy at the site of ingest or they may ingest content once from a staging server and designate the collection as closed, or no longer available for re-ingest. We will describe each of these scenarios and their preservation implications below.

Active Mode

By default, the LOCKSS software running on each cache in the network executes a routine but random crawling and polling mechanism in which it iteratively ingests content via HTTP or HTTPS using the base_URLs and crawl rules for a collection as recorded in the plugin and title database. Content ingested into a PLN using this active mode should remain intentionally accessible to the PLN network via the web. The active method allows the preservation network to catch file updates and, where crawling parameters permit, enables the addition of new files to the content base. It also enables caches that have corrupted content to repopulate that content using the authoritative source from the content contributor's site.

Closed Mode

In some cases, PLNs may choose to ingest collections that will not remain persistently available via the web. Content contributors may choose to avoid storing content that is embargoed or that is copyright protected on servers that are web accessible. In such cases, content contributors may move their content to a staging server temporarily for ingest into the network.[1] Once the content is successfully ingested and all designated caches agree that they have identical copies of that content, the content may be taken off line at the content contributor's site and the preservation copies become the authoritative versions of the content. If a file from such a collection becomes corrupt, the network will detect this corruption during its routine polling process. As the caches compare their copies, they will come to an agreement, in which they determine that a set number of caches match, and establish that the copy these caches hold is the authoritative version. One of these caches will then repair the corrupt copy on the inconsistent cache.

Re-crawl Intervals

For content ingested in active mode, the content contributor must set an appropriate re-crawl interval in their collection's plugin. Crawling any site (especially a large collection) more than approximately once a month tends to be too much overhead on a network and slows down polling and other preservation processes. Best practices currently state that the most frequently a static site should be re-crawled is once a month, and the least frequently a static website should be crawled is approximately once a year. Re-crawl intervals must also be set in the plugins for collections on closed collections—in that case being set to "never."

An additional implementation that is highly recommended for independent PLNs is to enable the Last_Modified configuration for the web servers hosting a content contributor's collections. This setting enables the LOCKSS daemon to compare information in the http headers of the archival units to determine if any files on the contributing site's server are younger than the file that the cache maintains on its local disk. The LOCKSS daemon can then ingest only those files. This reduces strain on the caches and the network as a whole.

CONTENT MONITORING

Content monitoring is a responsibility shared between the content contributor, each preservation site, and various designated central staff (either the LOCKSS team or a central staff hosted by the PLN). This section delineates the central staff and member institutions' responsibilities and describes the monitoring tools currently available to PLNs.

Staff and Member Responsibilities

Cache administrators at the content contributor site are responsible for ensuring that their content has been successfully ingested by all of the designated preservation sites. This requires the content contributor to not only verify replication, but also to perform both a proxy access of the content from a replicated cache site and a manual audit of the content.

Cache administrators at each preservation site are responsible for ingesting and monitoring content on their own caches. This entails ingesting new content when it is made available, monitoring the ingest process to make sure that it completes properly, and monitoring the communication that takes place between a cache and the other caches of the network during polling and repairing processes.

The central staff (PLN- or LOCKSS-based) bears the responsibility of administrating the overall network and monitoring its preservation activities. This administration includes monitoring the overall behavior of the network, watching the system logs, ensuring that polling and repairing processes are functioning correctly, and making sure that the network is running smoothly. For further details on network monitoring see Chapter 7: Cache and Network Administration for PLNs.

Monitoring Tools

There are currently three tools that assist in the content monitoring process: the cache manager, the LOCKSS daemon user interface, and cron reports:

- **The cache manager** is an open source, web-based tool co-developed by the LOCKSS team and the MetaArchive Cooperative. It queries the LOCKSS daemon on the individual caches and gathers status information concerning where AUs are being preserved, the overall status of the network, and cache storage capacity.

- **The LOCKSS daemon user interface** is a core tool provided by the LOCKSS software installation to all members of a PLN administering a cache. The interface lists the archival units available to the cache for ingest, communicates further information about storage capacity information, and serves as the primary tool for verifying the integrity of individual archival units, and initiating a restoration of content in the event of loss.

- **Cron jobs** can be configured to provide scheduled reports concerning the results of polling by the

LOCKSS daemon, and the status of archival units on the caches.

A cache administrator at a content contributor site can verify a successful ingest via the cache manager by observing which geographically dispersed sites have replicated its collection. The PLN member administering a cache, as well as any central PLN staff, can also monitor cache availability for ingesting new collections through the cache manager.

Through the LOCKSS daemon user interface, a cache administrator at a content contributor site or any central PLN staff member may also audit content at regular intervals through a proxy feature. This feature helps make certain that the harvesting parameters specified in the plugin continue to guide the LOCKSS caches correctly as they ingest and update content. Proxy collections from a cache become viewable from their base_URL in a standard web browser and can be manually viewed for their completeness and correctness.

Finally, central PLN staff members can initiate cron jobs to run automatically at a certain time or date and report on the results of polls conducted by the LOCKSS daemon between the caches. This communicates real time information about the health of the network and the content on the caches. These reports can be configured for delivery via email to the necessary staff and if so desired to technical and administrative staff at the content contributor's site as well.

RESTORING CONTENT

As alluded to above, in the event of catastrophic loss at a content contributor site, restoration can take place from another cache containing replicated content. Similarly, in the event of catastrophic loss at a preservation site, content can be restored by a re-ingest at the content contributor site or any other cache in the network that contains the relevant collections.

Restoring a Content Contributor Site

Restoring content to a contributing institution that has suffered catastrophic loss to their original content involves retrieving its preserved copies from any of the other caches in the network that bore responsibility for replicating that content.

A contributing institution wanting to access preserved content in a cache uses a web browser to request one or more preserved URLs through a Proxy Auto-Configuration (PAC) file distribution accessible via the LOCKSS daemon user interface's Audit Proxy feature. When one initiates an extraction via this feature it is possible to retrieve the entire collection exactly as it was previously ingested and/or re-crawled. With these restored pieces in hand, and with reference to the title database entry for a collection, a member should be able to reconstitute their collections on a new or restored web server within a reasonable time frame.

For information on the importance of data wrangling collections prior to ingest to reduce the amount of time spent reconstructing lost collections, please see the section titled "Collection Management" in Chapter 5: Content Selection, Preparation, and Management.

Restoring a Cache

When a cache fails, the contributing institution establishes a new cache on the network. This new cache re-ingests collections from the sites that it was previously assigned to ingest. Collections that are no longer available via the web (i.e., those designated as "closed" as described above) may be restored to the cache from any of the other caches that host copies of that content.

Additional Considerations

The LOCKSS software ensures that additions and alterations to files that fall within the harvest parameters of a plugin will be captured by the network as separate versions of the individual files. There is no automated way for LOCKSS to know whether a change at the content contributor's site was authorized or not, so it treats all of the files that it collects as authoritative. When LOCKSS collects a file that has been ingested previously, it does not overwrite the existing file, but rather makes a new version—recognizing that a content contributor may eventually desire to recover a previous version that reflects a more authoritative rendition of the content.

These versions are visible in a member's LOCKSS daemon user interface. If the content contributor experiences a catastrophic event that necessitates restoration using the preservation copies, all stored versions are easily accessible via this interface. The LOCKSS team is currently developing a secure implementation of a user interface that will allow access to the preserved content per the contributor's choice of dated versions.

CONCLUSION

Once content is successfully ingested into a PLN, all members must fulfill their individual and shared responsibilities to guarantee that content is being preserved. The goal of distributed digital preservation is to provide recovered authoritative content in the event of loss. To do so in the context of a PLN requires active involvement of the content contributor, the cache administrators, and the central staff during each phase of the preservation process.

ENDNOTES

1. A staging server is a dedicated server in which contributing members can temporarily store and organize collections that need to be harvested.

Chapter 7: Cache and Network Administration for PLNs

Matt Schultz (Educopia Institute)

Bill Robbins (Educopia Institute)

OVERVIEW

As covered in previous chapters, a Private LOCKSS Network (PLN) may rely upon the LOCKSS program for its network management and support or it may operate as a locally administered entity. This chapter begins by briefly touching upon the technical and administrative reasons why organizations might choose one approach over the other and then provides a technical overview of best practices in cache and network administration in both cases. The chapter closes with recommendations for effective communication tools and strategies that may assist PLNs in their cache/network administration activities.

LOCALLY ADMINISTERED PLNS VERSUS STANFORD-ADMINISTERED PLNS

All PLNs, by definition, rely on the LOCKSS software for their technical preservation infrastructure. The LOCKSS code is open source, but does not depend solely on the open source programming community to sustain its software development and maintenance activities. LOCKSS instead pioneered a sustainability model that requires that institutions that use the LOCKSS software, and that wish to receive support from the Stanford University-based team, pay annual membership fees to the LOCKSS Alliance. These fees support the LOCKSS central staff as they continue to update and refine the LOCKSS software. All PLN members, like the public LOCKSS network members, share in this relationship to the LOCKSS Alliance.

PLNs may adopt one of two approaches to administering and monitoring their networks: they may form as Stanford-administered PLNs or as locally administered PLNs.

Stanford-administered PLNs rely on the central infrastructure (title database, plugin repository, plugin development and cache manager) of the LOCKSS program. PLN staff are still responsible for activities such as content identification and content preparation. Stanford-administered PLNs rely on the program's workflows for testing and publicizing plugins to their repository, and for making content available through their title database. The LOCKSS team can also help Stanford-administered PLN members with tasks such as adding/removing caches.

Locally administered PLNs depend on the LOCKSS software, but choose to manage their own network infrastructure in all other aspects. They administer their networks, which may include maintaining their own configuration settings in the form of a separate LOCKSS title database, developing their own procedures to configure caches, and providing one or more plugin repositories. They must have agreed upon procedures to add or revise plugins, and need to build monitoring infrastructure to ensure proper network/cache operation.

Both methods of creating and hosting a PLN have yielded strong and efficient preservation networks.

CREATING AND MAINTAINING A PLN

The best practices for creating and maintaining a PLN depend upon the goals of the institutions that partner together to implement a preservation network. As covered in Chapter 3: Technical Considerations for PLNs and Chapter 4: Organizational Considerations, the degree to which a PLN relies upon the LOCKSS program's infrastructure or chooses a locally administered PLN structure depends upon a mixture of technical and organizational decision-making.

As the case studies in those earlier chapters demonstrate, PLNs that anticipate barriers to maintaining the technical infrastructure at their member institutions have chosen to rely upon the LOCKSS program's infrastructure. PLNs that seek to establish a sustainable infrastructure that does not depend on the LOCKSS team for their survival and that have sufficient technical expertise both centrally and at individual member sites may gain greater flexibility in the way that their networks are configured and run by establishing an independent-PLN infrastructure. With these priorities come certain best practices for configuring the caches and network that form such a PLN.

Recommending Hardware

As described previously, a PLN is composed of at least seven preservation caches, each of which runs the LOCKSS software and all of which are connected to one another through a network configuration. Because the LOCKSS software provides a high level of redundancy and governs the preservation activities of a network, the actual hardware upon which caches run can be fairly basic and inexpensive across both the dependent and locally administered scenarios.

The LOCKSS team recommends that each cache in a PLN be either a low-cost PC running the LOCKSS software from a Live CD or a low-cost UNIX or Linux-based server running a package installation of the LOCKSS daemon software.

As mentioned in Chapter 3: Technical Considerations for PLNs, PLNs have handled decisions related to purchasing hardware and configuring system requirements on the basis of maintaining conformity across the caches in their network. The most important factor is that the CPU capability of a cache matches the disk capacity since LOCKSS daemons continually hash, vote, and poll their content concerning its completeness and correctness. The more content contained in a cache, the bigger the disk, the faster the CPU needed for the LOCKSS daemon to perform its auditing activities. So long as each cache in the network can communicate efficiently with the other caches and are not bottlenecked by mismatches in CPU speed and disk space, the preservation network may select any hardware (both at the cache and network level) that it wishes.

LOCKSS Software Implementation

The choice of a Live CD (OpenBSD) versus a Linux (RPM) package installation of the LOCKSS software is largely a matter of technical and organizational priorities. A Live CD installation includes not just the LOCKSS daemon, but also a pre-configured, highly secure operating system. This type of installation is ideal for PLNs that seek to make it as easy as possible for their members to bring up and update their caches. This minimizes their individual members' technical investments.

For those PLNs that seek to establish themselves independently from the LOCKSS program, a Linux RPM-based package installation may be more appropriate. This has enabled such PLNs as the MetaArchive Cooperative to have broader flexibility with issues of hardware choices, security configurations, and content monitoring tool development/implementation for their preservation network.

Once the amount of content in a cache reaches a certain high quantity, the Live CD (OpenBSD) implementation becomes less optimal as Linux systems can handle large disk arrays more efficiently than OpenBSD.[2] As previously discussed in Chapter 3, the LOCKSS team expects to transition from the Live CD to a VMware virtual appliance in the near future.

Testing the LOCKSS Daemon

The LOCKSS central staff distributes regular daemon updates on behalf of both Stanford-administered PLNs and locally administered PLNs. For PLNs that use the Live CD (OpenBSD) installation there is an automatic software update system that upgrades caches automatically and securely. When the LOCKSS central staff releases a new daemon (approximately every six weeks), this software is automatically installed with no human intervention. Twice a year, the LOCKSS team releases a new Live CD. At that point, the cache administrator simply burns the new image to disk or drive, swaps them out, and reboots.

For PLNs using LOCKSS on Linux and implementing their own instance of the title database, updates must be completed manually. Current best practices suggest that a designated member institution or central systems administrator test each new release on a test network prior to its implementation on the production caches to ensure compatibility.

Test Network

The test network is a separate series of LOCKSS-configured caches that have been designated for testing, rather than production, purposes. It enables a PLN to run disaster scenarios and test cache repopulation. It also enables the PLN to test a harvest of each collection's content, which ensures that the LOCKSS daemon can successfully harvest those collections from a given member's hosted content. This also provides an imperative test of the accuracy of the plugin and manifest page created for each collection's content prior to its implementation in the production network.

The LOCKSS team maintains a test network for testing daemon releases and ensuring that submitted plugins can be used successfully to harvest collections at a PLN member site before they are made available to the PLN's broader production network.

Locally administered PLNs may establish and maintain their own test networks, which can consist of a single cache or a small network of two to three caches.

CONFIGURING A CACHE FOR THE NETWORK

Each cache in the preservation network must be configured to communicate with other peer caches in the network, as well as a central administrative server(s) where tools such as the title database, plugin repository, and tools that monitor the network as a whole reside. This section covers best practices and outlines steps for configuring a cache in both Stanford-administered and locally administered scenarios.

Cache Communication Configuration

When establishing any PLN, ensuring successful communication between all caches in the network requires only a handful of basic configuration settings. These settings will likely include such things as the:

- cache ip address and a fully qualified domain name
- subnet containing the IP addresses of local workstations that will need to access the LOCKSS cache's user interface
- mail relay information
- email address of the local system administrator
- user name and password for the LOCKSS daemon's web user interface.

The first two pieces of information are the most crucial, because they identify the participating caches and local workstations in the network's shared title database, which the LOCKSS daemon uses to carry out its preservation function and provide user feedback.

In a Stanford-administered PLN, configuring a cache includes choosing hardware that is compatible with LOCKSS recommendations, and collecting and following the detailed instructions for providing the information listed above during the software installation and initialization process. This process should be straightforward for any system administrator, and does not require any detailed knowledge of a UNIX operating system.

In a locally administered PLN, the set-up procedure varies depending upon what hardware and operating system are chosen. Depending on the level of automation for the centrally provided install and the amount of deviation of a cache's operating system and hardware from LOCKSS-supported platforms, configuration may be as simple as in a Stanford-administered network, or it may require more local expertise.

Further best practices call for no hard limits to be set on space allocated to any particular members' content. This allows for greater flexibility when decisions are made on content ingestion.

In the case of the MetaArchive Cooperative, caches that use the same hardware are configured with the exact same file system structure. Central staff provide member institutions with a kickstart file as a new round of caches are added to the network. The kickstart files function very much like the LOCKSS installer in that they require the same basic localized settings for installation. The result of kickstarting is a cache configured with security enhanced LINUX instead of OpenBSD, along with RPM package management and a LOCKSS daemon that makes use of SSL encrypted communication.

Secure Cache Communication

The LOCKSS inter-cache communication protocol (LCAP) has been designed to be resistant to attack even when used in an open network.[2] PLNs may choose to further secure their network by enabling the use of SSL. In this case, all network communication is both encrypted and authenticated. Encryption ensures that none of the data is visible to outsiders, even if the network infrastructure itself (routers, etc.) has been compromised. Authentication provides assurance that each cache is actually who it says it is—communication is allowed only among caches that possess cryptographic keys attesting to their identity. This option requires some additional actions by the PLN administrators to create and securely distribute cryptographic keys to each of the participating sites, and to update those keys as sites are added or dropped (see Communication Best Practices below).

CONTENT SERVER CONFIGURATION

Once a cache has been properly configured to communicate both with its peer caches and with the central administrative server, it is able to make use of the LOCKSS daemon to begin to crawl and ingest content from the web servers which host collections (for a thorough discussion of selecting and preparing content for ingest into the preservation network see Chapter 5: Content Selection, Preparation, and Management). Web server administrators have to ensure that all caches in the PLN have access to the slated content. If content is publicly accessible, web server administrators do not need to take any action. If content is access restricted, access must be opened. In order for the LOCKSS daemon to gain access to this content, local systems administrators have to add the IP addresses of all the caches to their network security settings, such as their web server allow lists, and firewalls.

Both types of PLN must have effective and consistent procedures to make these IP addresses known to all members of the PLN. If this is not done in a coordinated fashion and with consideration for the web server administrators at member institutions, content will be unavailable to some or all caches and the LOCKSS daemon will not be able to ingest, update, and preserve content.

The MetaArchive Cooperative, has streamlined this process by maintaining up-to-date IP lists, and supplying web server administrators with Apache configuration files to help automate the update process.

NETWORK ADMINISTRATION

As caches successfully engage in the LOCKSS system's preservation operations, there are a number of activities that must be carried out to ensure the proper maintenance of the preservation network as a whole. This section details those activities within the context of both Stanford-administered PLNs and locally administered PLNs.

Maintaining Network Configurations

The title database on the central administrative server contains XML formatted parameters that tell the LOCKSS daemon three crucial things:

1. location of plugins as signed jar files,
2. location and definitions of archival units, and
3. list of IP addresses for caches participating in a network (both test and production). Note: This is optional. A seed list may be used, which does not have to be updated for each new cache added to the network.

In both Stanford-administered and locally administered networks, the title database and plugin repositories are web hosted. Upon startup, the URL of the title database is passed to the LOCKSS daemon residing on each cache. Daemons extract the URLs of plugin repositories from the title database. A PLN may be setup such that the title database and plugins are available at the same web server. Alternatively a web server may simply host the title database, and plugin repositories may be spread over web servers maintained by member institutions. While open access to plugin repositories poses no problem, it is advisable to keep access to the title database restricted. Although caches are not easily compromised there is no reason to expose their IP addresses.

PLNs that rely upon the LOCKSS team for their title database must coordinate with the LOCKSS staff to ensure that this information is properly configured.

Locally administered PLNs must develop their own procedures for making sure that the title database and plugin repositories are properly maintained, as lack of updated information can impact the performance of the LOCKSS daemon as it carries out its myriad procedures. The primary concerns are that well-developed procedures are in place to ensure that:

1. Caches that join and leave the network are properly accounted for (optional);
2. New plugins or changes to plugins are easily and consistently propagated to the plugin repository; and
3. Newly available content/ AU definitions become part of the title database, so that network caches can learn about the new content.

Network Monitoring Tools

There are several monitoring tools available to PLNs, including the LOCKSS daemon user interface, the cache manager, and the implementation of cron jobs.

LOCKSS Daemon User Interface

The LOCKSS daemon user interface is a web-based tool provided by the LOCKSS installation. Local system administrators and central staff can log into the web based interface of a particular cache to view uptime, available and used disk space, the status of archival units that are already preserved on the cache, their size, files contained, and last crawl information. The interface also gives information about ongoing and completed votes and polls. LOCKSS daemons trace their action in a log file, which can be accessed via the user interface as well. In short, the interface provides detailed information and serves as the primary tool for verifying the integrity of individual archival units as well as the integrity of the cache.

Most importantly, the user interface can provide detailed information about crawl problems in cases where a LOCKSS daemon crawl may have been unable to reach the content in an archival unit. Standard network management best practices call for members to consistently alert the PLN staff of such issues. This leads to quick and consistent troubleshooting which helps the network to maintain a high rate of successful polling and voting.

Cache Manager

The cache manager is an open source, web-based tool co-developed by the LOCKSS team and the MetaArchive Cooperative. It queries the LOCKSS daemon on the individual caches and gathers status information concerning where AUs are being preserved, the overall status of the network, and cache storage capacity.

The cache manager assists by communicating to both central and local systems (or cache) administrators, such as the number of caches successfully participating in the LOCKSS daemon's voting and polling procedures, and which caches may be exceeding a storage threshold that can contribute to bottlenecks in the crawling procedures.

Cron Messages

PLNs may implement retrieval scripts that can communicate tailored information about the network's activities to suit their preservation priorities. For example, cron jobs (automated Linux commands or shell scripts) can be scheduled to run automatically at specified times and dates. Locally administered PLNs can configure cron jobs to provide scheduled reports with details concerning the results of polling by the LOCKSS daemon, the status of archival units on the caches, and changes that have been made to resources and directories on the administrative server.

The following are examples of information that can be scheduled for such reports, as exemplified by the MetaArchive Cooperative's use of this mechanism:

- Completed polls: incidence of success in securing a sufficient number of caches to reach agreement on a request for vote; as well as completion of the vote—expressed as a quantity
- No quorum errors: incidence of failure to secure a sufficient number of caches to participate in a poll—expressed as a quantity
- Fetch errors: incidence of a failed crawl/re-crawl of an AU by the LOCKSS daemon—identifies the AU in question
- Can't fetch permission page error: incidence of failure of the crawl process to fetch the manifest or permission page of a collection—identifies the AU in question
- Low replication warning: communicating that an AU has not been effectively distributed across a sufficient number of caches—identifies the AU in question

Managing and Making Effective Use of the Tools

The tools can identify problems, but staff members must ultimately take steps to resolve them. Also, whatever tools a network uses will require maintenance. The LOCKSS team takes care of these responsibilities for Stanford-administered PLNs.

Locally administered PLNs must determine who is responsible for setting up and maintaining network monitoring tools and assign responsible software engineers to resolve problems that are discovered with the monitoring tools. Crawl problems reported by caches trying to access content will likely need to be resolved by local web server administrators. Plugins that fail to crawl content will need to be fixed by plugin developers. Caches that do not respond may have to be rebooted by local cache administrators. These resolutions all need to be assigned and prioritized effectively in a locally administered PLN.

Administrative Server

The title database and plugin repositories are hosted on an administrative server(s). Whether a PLN uses the cache manager or another network monitoring tool that may also reside on this server(s), it is preferable to set it up in an environment that is not widely accessible. Access to the title database via the Internet should be restricted to PLN network caches and PLN members. For example, the MetaArchive Cooperative has chosen to host the cache manager, title database, and its central plugin repository on a single server, which they recently migrated to the Amazon EC2 Cloud. Few users have access to this server.

The LOCKSS team takes care of these considerations on behalf of Stanford-administered PLNs. Locally administered PLNs need to consider where to locate these tools, configuration files, and attendant scripts to update components.

Any network monitoring tool needs to query LOCKSS caches about their status. Thus caches must allow access from the server that hosts the monitoring software. System administrators for PLN caches need to make sure that their local network settings allow access to the LOCKSS daemon user interface port. Which port a LOCKSS daemon uses is part of its local configuration. In Stanford-administered PLNs, this location is set by the LOCKSS team. Locally administered PLNs should agree on a common port number used by all caches.

Best practices for locally administered PLNs recommend that an administrative server(s) is not placed under the control of any single member site, as this makes the network dependent on that member's continued participation. Instead it should be administered in a location central to that PLN. For the MetaArchive Cooperative, as mentioned, this has led to the use of Amazon EC2 cloud servers.

Administrative Server Backups

A PLN should ensure that central administrative components are not bound to a single member institution. This carries over into backup strategies as well. Two strategies for backing up these resources that may be of some appeal to locally administered PLNs are:

1. To configure a backup server in a cloud environment
2. To partner with a separate preservation network (PLN or other form) to host a backup server

The MetaArchive Cooperative, as a locally administered PLN, is pursuing both strategies through their use of Amazon EC2 and through their collaboration with the San Diego Supercomputer Center through the NDIIPP-sponsored Chronopolis Project, which hosts a distributed digital preservation network that runs the Storage Resource Broker (SRB) infrastructure.

COMMUNICATION BEST PRACTICES

Distributed digital preservation, by definition, requires communication and collaboration across multiple locations and between numerous staff. Success in establishing and maintaining a PLN requires timely coordination between member institutions and any central staff that may exist to help monitor and maintain the network.

Documentation

Documenting the development and decision-making that takes place in the context of establishing and maintaining a PLN, regardless of its technical and administrative organization, is essential, since digital preservation solutions succeed in part on the basis of accessibility, transparency, and accountability. Some recommendations for hosting working copies, as well as completed versions, of documentation for PLN members and the interested public are:

- **Wikis:** wiki software (e.g., MediaWiki) allows for collaboration on document development, and can be configured for both internal and public publishing
- **Public Websites:** websites can provide a hosted outlet for authoritative and guiding documentation for the PLN

- **Content Management Systems:** robust and modular web development software (e.g., Drupal) can provide various levels of user permissions for managing information of importance to the PLN

Announcements and Alerts

Ensuring that technical and administrative staff at both partner sites and any central administrative location receive timely alerts about problem issues in the network can make or break a PLN. Announcements that impact the organization must go out in a timely and, occasionally, secure fashion to solidify decision-making and network policy implementation. Some recommendations for ensuring such timely secure communication are:

- **Dedicated Listservs:** listservs can be managed by a partner site, or, if cost effective, hosted through a service; listservs should be targeted toward a PLN's organization specifics (e.g., a technical listserv, a administrative listserv, and perhaps a general communication listserv, etc.).
- **Ticketing Systems:** project management and bug/issue tracking systems (e.g., Trac) can enable the technical staff at partner sites to solicit feedback, and any central staff to prioritize issues and responses.
- **Conference Calls:** regularly scheduled calls between central staff and technical and administrative staff at partner sites assists in deliberation on PLN-wide issues of concern.
- **Encryption Keys:** these allow files and messages to be encrypted and decrypted for electronic transfer across a public network. Keys may be helpful when contributing members in a PLN need to send sensitive information like logins/passwords to one another. Encryption keys are best distributed and delivered in-person or via postal mail on write-protected media.

Meetings and Workshops

Prolonged human endeavors have never been successful without some level of face-to-face interaction and interfacing. To ensure that a PLN is successful in catalyzing the collective knowledge and expertise of its member institutions, it is helpful to sponsor in-person meetings:

- **Annual Meetings:** these provide opportunities for the administrative heads of partner institutions to engage in vital decision-making, and to solidify preservation planning as the PLN moves forward.
- **Workshops:** these provide for the transfer of technical knowledge and expertise across the members of a PLN, allow for focused problem solving, and solidify best practices.

CONCLUSION

The degree to which a PLN chooses to rely upon the LOCKSS infrastructure or establish itself independently is a key distinction between PLN types that results in a different set of administrative needs/choices. With each approach comes a set of concerns and responsibilities on the level of cache and network administration. These should be carefully weighed by institutions that seek to host a PLN.

ENDNOTES

1. The relationship between the quantity of archival units and the amount of disk space consumed in the context of necessitating a shift away from the Live CD implementation of the LOCKSS software is somewhat of a moving target; PLNs should consult the LOCKSS team if they require clarification.
2. See: http://www.eecs.harvard.edu/~mema/publications/hotel.pdf (last accessed 12-21-2009).

Chapter 8:
Copyright Practice in the DDP: Practice makes Perfect (or at least workable)

Dwayne K. Buttler (University of Louisville)

OVERVIEW

Copyright limits the use of many creative works, including some of those destined for distributed digital preservation (DDP) initiatives that use Private LOCKSS Networks (PLNs) or other technology solutions. The copyright within the creative works governs how they might be copied, distributed, and hence ultimately preserved. In theory, copyright law is simple but in practice far more difficult.

Copyright can exert a cumulative effect on a DDP. A DDP locates multiple digital copies of a work in a geographically distributed fashion in order to best safeguard the work. Each copy and distribution in copyright law is a potential infringement, and raises liability for possibly infringing reproduction and distributions rights. The more copies reproduced and distributed, the greater the potential liability.

This multiplying effect is intrinsic to DDPs. However, contributors and managers of DDPs can overcome this cumulative effect if institutions effectively manage copyright from the earliest stages of contributing content in a DDP. The cumulative effect more accurately echoes the dated underpinnings of pre-digital copyright law rather than practical barriers to creating DDPs.

ORIGIN AND SCOPE OF COPYRIGHT

Copyright law is territorial in both its origin and scope. Many nations have distinct copyright laws. Some similarities exist among those laws due to international treaties and other arrangements; however, vast differences also separate them. The fair use doctrine, for instance, is unique to United States copyright law.

This chapter specifically addresses United States copyright law. Contributors and entities outside of the U.S. who are contemplating either joining or creating a DDP also need to understand and manage concerns arising under laws governing copyrights and limitations in their countries of origin, and, in some cases, the DDP's country of origin as well.

The scope of US copyright law is broad and disturbingly ubiquitous. Copyright in the U.S. automatically protects original expression that is fixed in a tangible medium of expression. Consequently, even seemingly mundane artifacts such as email messages might rise to the threshold of protectability (i.e. "originality" in U.S. law), and will enjoy federal copyright protection at the moment an author fixes the email (or other work) electronically or by other means. The threshold for protection is that low, and the barriers to acquiring copyright are that few.

The current term of protection is the life of the author plus 75 years. Accordingly, some (or many) materials dating from the early 20th century may still enjoy copyright protection, subject only to having earlier satisfied some modest conditions. This excessive length of protection, coupled with misunderstandings about copyright, can have unintended (or perhaps carefully cultivated) consequences for preservation—many materials well worth preserving stay at risk because of real or perceived limits in the law protecting the real or imagined economic market for the work, not due to any technological barriers that limit digital preservation.

The exclusive rights of copyright would seem to limit, or even completely prevent, preserving works by the reproduction and distribution of them. However, exceptions such as the fair use doctrine in U.S. law, or contractual agreements, such as deeds of gift or similar deposit instruments and licenses to use copyrighted works, frequently temper that exclusivity and thus lessen or eliminate potential infringement and liability possibilities.

DDPS AND COPYRIGHT

Preserving materials using DDP strategies and technologies involves three sometimes distinct, but often overlapping, stages of effort: contributing, preserving, and retrieving content. The contributing stage requires the contributor and contributing institution to closely examine content prior to its submission to the DDP. The contributor holds sole responsibility for resolving copyright concerns in contributing content as part of its submission policies and procedures. The preserving stage focuses on the relationship of the initiative to copyright law and the administration of contributed content, legal, business relationships, and DDP operations among its various contributors. The retrieval stage implicates decision-making within the DDP's organizational structure and beyond, raising questions around issues of access and use after materials have been submitted to the DDP system or repository. Each of these stages has sometimes unique, sometimes overlapping, concerns, as illustrated in the following sections.

CONTRIBUTING CONTENT

This stage in particular requires the contributor and contributing organization to closely examine each potential item of content. Not surprisingly, the contributor must first identify the individual copyright concerns and then resolve them. This effort is much easier said than done, given that contributing content can raise many thorny copyright questions. This stage deals fundamentally with clarifying copyright and complying with copyright law by developing and using reasonable and good faith strategies.

One prong of this analysis, and an essential strategy, addresses copyright proper: the federal law, the exclusive rights, its requirements, and limitations. Contributors must determine (or strive to determine in some cases) whether the work is in fact protected by copyright in order to understand the copyright implications of using the work. Determining whether a work is protected by copyright requires undertaking an assessment of the provenance of the work, including investigating facts and circumstances surrounding its creation and subsequent developments occurring throughout its history.

Important questions in this analysis typically would include:

1. When was the work created?
2. Under what circumstances?
3. Who created the work?
4. Was it created in the United States?
5. Was it registered with the U.S. Copyright Office?
6. Does the physical artifact bear a copyright notice (the familiar c in a circle © or copyright year name information)?

7. Is the artifact a visual work? Print material? Audio or visual? Computer software?

8. Was the creator of the work acting as an employee for an organization when he or she created the work?

9. Was the work published in a copyright law sense (which can differ greatly from the common lay understanding of publish)?

10. Is the work a sound recording which may enjoy state law protection similar to federal copyright protection but for a longer period?

Equally important, when the work is indeed protected by copyright, the contributor must then resolve how to use it lawfully. One important possibility for lawful use is to secure permission or license.

An essential facet of copyright ownership is the power to grant another party rights to make use of all or parts of the copyrighted work in broad or carefully proscribed ways. Copyright owners can give permission to make use of protected works, and may carefully define the scope and extent of that permission in a written (sometimes oral) agreement. Undertaking a permission analysis, particularly in the library and preservation communities, requires contributors to examine not only current organizational practices, but also to uncover past organizational actions that would have governed gifts and donations of special materials, many of which may still enjoy copyright protection, yet would benefit greatly from DDP safeguards.

Deeds of gift or deposit instruments often accompany deposits of such materials. Practices and language in those instruments vary widely. Some instruments may assign copyright to the organization receiving the materials. Other instruments might license specified uses of the materials for an agreed upon time or even indefinitely. Still other instruments might serve as contracts between the depositor and the library, setting forth specific and ongoing obligations for the library and the donor.

The particulars of each of these instruments are often distinct from one another, but they can also often overlap. The key to interpreting these instruments is a close reading with reference to their legal tradition. Some deposit instruments act principally and clearly as assignments of all right, title, and interests in the works, thereby making the library the owner of both the works themselves and the rights held within them.

This outcome is ostensibly clear. On the other hand, a grantor can only grant those rights that are genuinely owned by the grantor. Thus, even in deposit instruments that purport to give all rights, title, and interest to the library, that statement of seeming transfer is accurate only to the extent that the depositor owns such rights.

Moreover, in some cases, the lack of a specific reference to copyright might prove detrimental, if not fatal, to the assignment. Some courts have zealously required the word *copyright* to appear in an assignment to ensure that the person making it –clearly understood that *copyright* itself was transferred, not solely ownership of the physical artifact. This decision making from the courts may reflect a minority position in copyright law but still influences some federal circuits. That principle would also apply to licenses or other conditional arrangements specified in a deposit instrument that purportedly grants rights that the grantor does not own and, therefore, cannot give to the institution.

Permission may also include identifying and contacting copyright owners in order to secure permission retroactively, or even proactively, for materials destined for the DDP. Here again understanding provenance is essential to securing permission. Unfortunately, in many situations, crucial underlying facts and circumstances are simply missing from the historical record, particularly as the time between the creation and use of the work increases. Given that works dating from 1923 to today might still enjoy copyright protection, gaps in the historical record seem certain for many works of cultural memory. Securing permission to use these works is at best unlikely, and, in many cases, not possible at all.

Contributing protected works to a DDP will require the analysis and application of lawful copyright limitations to the works. The sole other option is to not preserve them at all. Foremost among limitations in the law is the fair use doctrine outlined in Section 107 of the U.S. Copyright Act of 1976. In many circumstances, understanding and weighing the four factors of fair use—purpose, nature, amount, market effect—will become essential to including materials in a DDP.

Fair use forms the bedrock for allowing many uses of copyrighted works, and, in particular, strives to support activities that bring new social value to protected works, such as long-term preservation for future generations. Analyzing fair use necessitates an appreciation of the array of legal cases that have interpreted the statutory language of Section107. In general, the Supreme Court and judges in Federal District and Courts of Appeal have clearly favored *purposes* that enrich the public good while simultaneously leaving the *market* for the original copyrighted work economically unharmed.

This emphasis on purpose and market effect has come to dominate fair use analysis, leaving courts with many opportunities to discount the remaining two fair use factors: amount and nature. In the DDP context, this emphasis on market effect is likely to favor long-term preservation opportunities, as they enrich the public good and exert no or limited effect on the market for the original work. The dark archive approach to digital preservation may support findings of no or little market effect given that only custodians and original contributors may access the preserved works. The multiple copies of works within a DDP exist to ensure recoverability and not to provide broad accessibility.

Moreover, some works of interest to DDP contributors have at best thorny trails of provenance, and often are simply orphans of copyright. These artifacts of a law gone awry, coupled with excessive duration, reflect little possibility for engaging an ongoing market, if ever they had a viable market at all. On the other hand, these same works afford a glimpse into the time and place of a culture adjusting to new technologies and communication media, while simultaneously using extant media to describe these changes. They form the core vehicles of particular value to future explorers of history and culture. Preserving them is fundamental to preserving cultural memory.

Preservation is also a key rationale for the library reproduction opportunities found in Section 108 of copyright law. Section 108 has well served the library community in achieving important missions since its inclusion within the Copyright Act of 1976. This section generally brings greater certainty to some routine library activities that would otherwise most *likely*, but not most certainly, establish fair use under Section 107.

However, it brings little influence and opportunity to digital preservation initiatives writ large. Limitations on the dissemination of digital copies beyond the *premises* of the library or archives eviscerated Section 108 in 1998 and hamper its meaningful use in support of DDP activities.

The copies in a DDP are by definition digital, and by purpose distributed beyond the premises of the holding library. The current language of 108, however, limits digital copies to the premises of the library. This limit reflects a core, albeit flawed, linkage for some owners of copyrighted works: digital copies equal clone copies and clone copies equal widespread dissemination.

Section 108 may have some limited application for preserving unpublished works that are not otherwise distributed in that digital form. However, this language and possible support is less than clear in meaning, and nebulous in application within a DDP. Is it possible to segregate each work within that category as a decision-making and policy choice? This consideration would in turn lead us back yet again to look closely at appropriate fair use strategies for supporting ingest of materials into the DDP.

PRESERVING CONTENT

Copying and distributing content for preservation is the next stage in the technical process. This stage more specifically implicates decision making in the realm of the distributed network itself and the broader organizational framework. Careful analysis and responsible decision making are required to determine how, and to what extent, preservation is possible in light of existing law, and then to identify possible strategies for using digital technology. Decision making will focus on the relationship of the DDP to copyright law and on the administration of contributed content; legal and business relationships; and DDP operations among the DDP, contributors, contributing organizations, affiliates, and funding agencies.

Contributors can substantially lessen many overarching concerns by closely scrutinizing in-house management practices during the ingest stage. However, the DDP as an organization must still exercise vigilance in creating administrative, organizational, and technological controls and strategies for governing access to DDP materials. One possible control is restricting access to a specific collection to only the contributor of that collection and the necessary DDP custodians, particularly in those cases where the contribution is grounded in fair use, and not specified in a license or deeds of gift. In other words, the DDP should generally function as an archive, and not as a content provider, in order to bolster fair use and other opportunities of use not adequately supported in larger distribution contexts.

Some questions arising in the preservation stage would typically include:

1. Has the organization designed and implemented standards by which contributors can fulfill their obligations and responsibilities for managing copyright and accepting responsibility for that decision-making? How does the DDP manage and address ownership of software, technologies, and other works resulting from contributors or combined efforts of multiple contributors?

2. How does the DDP address any requests to modify the standard agreement language and assure compliance with copyright conditions and standards?

3. When might infringement liability attach to the organization and how has the organization managed that legal and business risk?

4. Who can access all or only limited content and what are the security standards for acting as a custodian? How is access defined, and under what conditions?

5. What technological controls are needed and deployed to make reasonable assurances that the uses are limited to only relevant users or in some sense to assure the appropriateness and darkness of the archive?

6. Who determines possible sanctions for agreement violations and how are those assessed?

7. Has the DDP acquired sufficient rights from contributors to make multiple copies of the works and to satisfy its legal obligations?

8. Does access equate to use under copyright law and what direct infringement could support secondary infringement liability for the DDP or other contributors?

9. Are authentication measures in place to substantiate access controls in a fashion consistent with reasonable practices?

10. What rights does the DDP hold to permit other contributors to make copies in geographical locations in the country of origin or elsewhere?

Successful management of copyright concerns at the preservation stage correlates directly with effective management of copyright at the ingest stage. Related considerations for managing copyright across the whole of a DDP emphasize organizational decision-making as an institution. The interests among the DDP, contributors, and contributing institutions may not always align and concerns about copyright uses and potential liability might magnify those divergent interests.

Resolving these potential differences raises additional concerns about how copyright law influences the organization and contributors. It also helps give substance to the DDP's perspective on copyright and signals to external parties the value the DDP places on managing and respecting the rights of copyright owners. The MetaArchive Cooperative is one example of a DDP that has carefully documented the responsibilities of contributing institutions and the roles and responsibilities incurred by all members regarding copyrighted content. The MetaArchive Cooperative's charter[1] and membership agreement[2] deal explicitly with member responsibilities and require each contributing institution to agree to abide by U.S. copyright law, acknowledge that they hold sufficient rights or licenses to contribute content to a multi-site preservation initiative, and to hold other members harmless in the event of infringement. Upon signature of the membership agreement members agree to a framework of trust that is paramount to the success of a DDP initiative.

RETRIEVING CONTENT

Retrieving content from the DDP is the final process that requires copyright analysis. Analysis at this stage also implicates decision making within the DDP structure and possibly outside of it, raising questions not only from contributors seeking access to materials which they did not submit, but also from third parties unrelated to the DDP seeking access to materials. Retrieving content also links closely to the copyright status of each resource in the DDP. As a general principle, if the contributor made reasonable steps to comply with copyright at the point of ingest, each contribution should be of less concern upon retrieval. That relationship brings the interests of contributors and the DDP into close alignment.

The *dark archive* approach to constructing some DDPs, as employed by the MetaArchive Cooperative, is emblematic of this access and use analysis and its copyright implications. A dark archive is a core manifestation of a copyright law focusing myopically on ardently protecting copyrighted works, rather than zealously securing their long-term availability and preservation.

Copyright law can support DDP initiatives, but managing copyright is necessary and fundamental in considering DDP contributions, development strategies, and related administrative and business decisions. Some of the content stored in the DDP might be lawful to use and disseminate in other channels; individual contributors may be lawfully providing access to some content through other means. In a dark archive, however, the DDP should isolate materials to preserve important foundations supporting at least those resources included under fair use.

Contributors should enjoy access solely to their contributions and not to content contributed by others. More importantly, even if alternative uses are lawful in other channels, given the vast number of contributions housed in a typical DDP, it is difficult to uniformly achieve the granularity of copyright understanding necessary to safely provide access for all of a DDP's contents. Achieving that granularity through mastering the inconsistencies of current U.S. and international law is at best unlikely, and in reality virtually impossible.

Consequently, this stage also emphasizes organizational decision-making, much of which shares similarities and overlap with the copyright analysis discussed in the previous section. Some questions in the retrieval stage might typically include:

1. What limits does the DDP place on access to materials and what management schemes support those limits?

2. Has the DDP developed and deployed clear standards for addressing copyright concerns between the DDP and contributors at each stage—ingest, preservation, and retrieval?

3. To what extent are the items that are retrieved copies of the original rather than new originals and therefore not copies in a copyright sense?

4. Does inclusion in a multi-stage, multi-copy environment affect whether the work is published

with permission from the owner (or without it) under the 1909 and 1976 Copyright Acts and what consequences arise if the DDP published it? Can publication occur at all in a dark archive?

5. Does copyright reside in the digitized copies, independent of the originals as a function of the digitization itself, and subsequently who owns that new work? Is there sufficient originality in digitization to support an independently copyrightable work?

6. Who can access content in the DDP and how is access isolated among multiple contributors and multiple access possibilities?

7. How does global and local access relate to the underlying analysis of fair use or deeds of gift undertaken at the ingest stage? Does fair use support global access for some items?

8. Can a deposit instrument that purports to convey a license to use a work survive the termination of copyright protection in the subject of the license itself?

9. What is the ongoing enforceability of a license in which the subject matter entered the public domain because of failures to renew it under the 1909 Act?

10. Does the DDP have ongoing rights after the withdrawal of contributed content or does the DDP have an obligation to withdraw content arising from potential challenges by the copyright owner and eliminate all remaining copies?

The retrieval stage, like most final activities, illuminates much of the circularity of copyright. It also highlights the unity of copyright and the DDP itself as an organizational concern and a business with risk. Viewing copyright (in the context of DDP) as a three-stage process offers a useful construct for sequencing copyright decision-making, but in reality each stage simply represents multiple facets of a singular goal—complying with copyright law.

Unfortunately, complying with copyright law embodies the essence of law and conversations about liability and risk. DDPs are inseparable from this conversation, and, in some sense, have introduced new questions of complexity. Risk is endemic in everyday life, and DDP strategies reflect that reality again and again. Copyright limits the making of copies in order to protect the market of the copyright owner. Making multiple copies therefore magnifies risk under traditional copyright precepts. But making multiple copies also substantially increases the likelihood that a particular work will survive today and into the future of the digital maelstrom we now occupy.

Traditional precepts emphasize the economic value of the work, but for many works in the DDP, economic value is intimately akin to their unique time in history and is hence long gone today. The preservation value and resulting social benefit is far higher than any speculative and remaining economic market. These countervailing factors weigh heavily in fair use analyses, but it is not clear how and when they will apply in all cases. The element of risk and the existence of thorny options are both inherent in fair use analysis. Erring on the side of not preserving raises the likelihood that the work will disappear from the historical record. Relying on fair use as a default strategy (permitting all inclusions in the DDP), even those operating strictly as dark archives, increases risk in a legal and business sense. Applying reasonable and thorough practices, particularly prior to the ingest of content into a DDP, including reaching good faith relationships among contributors concerning copyright throughout that content's preservation, can go a long way toward resolving this common conundrum.

CONCLUSION

Copyright and new technologies have a long history of conflict, tolerance, co-existence, and ultimately widespread adoption. DDP is ultimately a strategy and a business practice embracing new digital technologies in order to serve an age-old goal—keeping cultural memories alive for future generations, no matter the medium of original dissemination. Copyright law supports that goal in principle and practice; it just sometimes takes a lot of practice.

ENDNOTES

1. MetaArchive Charter: http://www.metaarchive.org/public/resources/charter_member/MetaArchive_Charter_2010.pdf (last accessed 1-31-2010).
2. MetaArchive Member Agreement: http://www.metaarchive.org/public/resources/charter_member/Membership_Agreement_2010.pdf (last accessed 1-31-2010).

Appendix A:
Private LOCKSS Networks

THE ALABAMA DIGITAL PRESERVATION NETWORK (ADPNET)

Profile

The Alabama Digital Preservation Network (ADPNet) is a membership organization governed by the Network of Alabama Academic Libraries (NAAL), a unit of the Alabama Commission on Higher Education in Montgomery, Alabama. ADPNet was established in 2006 with a two-year National Leadership Grant awarded by the Institute of Museum and Library Services (IMLS). It has been a self-sustaining network since 2008. Day-to-day management of ADPNet is conducted by the members themselves, through an elected Steering Committee. The NAAL Advisory Council exercises general oversight. ADPNet does not currently require a membership fee.

The mission of ADPNet is twofold: first, to manage and sustain a reliable, low-cost, low-maintenance preservation network for the long-term preservation of publicly available digital resources created by Alabama libraries, archives, and other cultural heritage organizations; and second, to serve as a model and resource to other states and consortia that are interested in setting up digital preservation networks of their own. ADPNet seeks to foster better understanding of distributed digital preservation methods in the state and to create a stable, geographically dispersed "dark archive" of digital content that can be drawn upon if necessary to restore collections at the participating institutions (listed below).

At present, harvested content consists primarily of archival audio, video, and still image files. Among the digital resources that have been harvested into ADPNet are the Alabama Department of Archives & History World War I Gold Star Database; the Auburn University Historical Maps Collection and Sesquicentennial Lecture Series; the Troy University Postcard Collection; the University of Alabama 1968 Student Government Association Emphasis Symposium, with sound files of historic speeches by Robert F. Kennedy, Ferenc Nagy, and John Kenneth Galbraith; the University of Alabama at Birmingham Oral History Collection;

and the University of North Alabama William McDonald and U.S. Nitrate Plant collections. The network plans to harvest several terabytes of new content in 2010.

Participating Institutions

Alabama Commission on Higher Education, Alabama Department of Archives & History, Auburn University, Spring Hill College, Troy University, University of Alabama, University of Alabama at Birmingham, University of North Alabama

Contact Information

Mr. Ron Leonard
Network of Alabama Academic Libraries
Alabama Commission on Higher Education
P.O. Box 302000
Montgomery, AL 36130-2000
Phone: (334) 242-2211
E-mail: ron.leonard@ache.alabama.gov

Mr. Aaron Trehub
Auburn University Libraries
Auburn University, AL 36849
Phone: (334) 844-1716
E-mail: trehuaj@auburn.edu

Website

http://www.adpn.org/

CLOCKSS (CONTROLLED LOCKSS)

Profile

CLOCKSS (Controlled LOCKSS) is a not for profit joint venture between the world's leading scholarly publishers and research libraries whose mission is to build a sustainable, geographically distributed dark archive with which to ensure the long-term survival of scholarly publications for the benefit of the greater global research community.

As libraries migrate to online-only subscriptions, they expect assurances from publishers that their shared investments are protected and preserved for generations to come. The CLOCKSS archive provides this assurance via Private LOCKSS Network technology. The CLOCKSS archive truly serves the world's scholars. Content preserved in the CLOCKSS archive is made freely available to all when it is not available from any publisher. Examples of publisher's content now free for all to use include the Oxford University Press's journal Brief Treatment and Crisis Intervention, and Sage Publications's journals Auto/Biography, and Graft.

Participating Publishers

http://www.clockss.org/clockss/Participating_Publishers

Participating Libraries

http://www.clockss.org/clockss/Participating_Libraries

Contact Information

Victoria Reich
CLOCKSS Archive
1450 Page Mill Road
Palo Alto, CA 94304
Phone: (650) 725 1134
E-mail: info@clockss.org

Website

http://www.clockss.org/clockss/Home

THE COUNCIL OF PRAIRIE AND PACIFIC UNIVERSITY LIBRARIES (COPPUL)

Profile

The Council of Prairie and Pacific University Libraries (COPPUL) is a consortium of 21 university libraries located in Manitoba, Saskatchewan, Alberta and British Columbia. Member libraries cooperate to enhance information services through resource sharing, collective purchasing, document delivery, and many other similar activities. COPPUL's vision is to be a cohesive and collaborative organization that provides leadership in the development of solutions that meet the academic information resource needs of its member institutions.

The COPPUL Private LOCKSS Network is a program that utilizes the LOCKSS digital preservation system as a means to archive collections of local interest to members of the Council of Prairie and Pacific University Libraries (COPPUL) that are not being preserved through any other means. Digital materials such as small university press publications, open access journals and other electronic journal collections, born digital government publications, and locally created digital collections that are at risk of being lost are preserved as part of the program.

Participating Institutions

Athabasca University, Simon Fraser University, University of Alberta, University of British Columbia, University of Calgary, University of Manitoba, University of Saskatchewan, University of Victoria, University of Winnipeg

Contact Information

Denise Koufogiannakis
Chair, Steering Committee
University of Alberta Edmonton, AB T6G 2J8
Phone: (780) 492-5331
E-mail: denise.koufogiannakis@ualberta.ca

Website

http://coppullockssgroup.pbworks.com/

THE DATA PRESERVATION ALLIANCE FOR THE SOCIAL SCIENCES (DATA-PASS)

Profile

The Data Preservation Alliance for the Social Sciences (Data-PASS) is a broad-based partnership devoted to identifying, acquiring and preserving data at-risk of being lost to the social science research community. Examples of at-risk data include opinion polls; voting records; large-scale surveys on family growth and income, social network data; government statistics and indices; GIS data measuring human activity; qualitative text, video, and audio records of subject interviews; and other data describing human behavior, society, and institutions. The partners coordinate identification, acquisition, and cataloging of data at risk; develop best practices for data archiving; and create a shared infrastructure and practices for cataloging and preservation.

Data-PASS provided a shared electronic catalog for the tens of thousands of studies or series that comprise all partners' entire data holdings. The Data-PASS shared catalog establishes, for the first time ever, a unified way to find social science digital data in major U.S. archives and comprises the world's largest catalog of social science datasets.

The Data-PASS partners (listed below) have built a prototype storage platform for distributed replication of digital holdings. The partners have received funding from the Institute of Museum and Library Services to continue development of this system and release it as an open source tools for libraries, museums, and archives that wish to collaborate in replicating their own content.

This prototype system is built around a core of Private LOCKSS Networks; a schema to encapsulate inter-archival replication commitments; an automated schema-driven service that audits PLN's; and Open Archives clients to harvest data collections from the Dataverse Network and other repositories using the Data Documentation Initiative (DDI) schema.[1] For more details, see the project page at: http://data-pass.org/syndicated-storage.jsp.

Participating Institutions

The Inter-university Consortium for Political and Social Research; the Roper Center for Public Opinion Research; the Howard W. Odum Institute for Research in Social Science; the Henry A.

Murray Research Archive, a member of the Institute of Quantitative Social Science; Electronic and Special Media Records Service Division, National Archives and Records Administration; the Harvard-MIT Data Center; the Library of Congress

Contact Information

Data-PASS
ICPSR
University of Michigan
Institute for Social Research
P.O. Box 1248
Ann Arbor, MI 48104-2321
Phone: (734) 763-6075
E-mail: data-pass@icpsr.umich.edu

Website

http://www.data-pass.org

ENDNOTES

1. Altman, M., Beecher, B., Crabtree, J., Andreev, L.,Bachman, E., Buchbinder, A., Burling, S., King, P., & Maynard, M. (2009). "A Prototype Platform for Policy-Based Archival Replication." *Against the Grain*, 21(2).

THE LOCKSS-KOPAL INTEROPERABILITY PROJECT

Profile

Scholarly information today is mainly born digital and is increasingly made available through digital means. This makes effective long-term digital preservation urgent for researchers in every field. Scholars know this is important, but generally assume that others, especially libraries, will address the problem. No simple solution exists. Multiple backups are a help, but do not address problems of integrity or authenticity or usability and older backups on tape are particularly vulnerable to physical or magnetic decay. Serious research on this topic has gone on for a decade with valuable tools such as LOCKSS and KOPAL as a result, but far more research remains to do. The problem of long term digital archiving is not solved, but becomes more solvable as the research moves forward. This project sets up a network infrastructure that takes an active step toward preserving bitstream integrity while ensuring readability, and then tests the product with materials from German institutional repositories.

Specifically this project proposes interoperability between the open-source elements of two existing archiving systems (LOCKSS and KOPAL) in order to combine cost-effective bitstream preservation with an established tool for usability maintenance and format migration.

Based on these goals the chief elements of this project are:

1. to establish a cost-effective LOCKSS network in Germany including infrastructure to provide ongoing technical support and management for LOCKSS and its variants (e.g. CLOCKSS);
2. to conceptualize and implement interoperability between LOCKSS and KOPAL in order to combine cost-effective bitstream preservation with well-developed usability preservation tools; and
3. to test the interoperability prototype by archiving data from German institutional repositories.

Participating Institutions

LOCKSS (Lots of Copies Keep Stuff Safe), The German National Library, Humboldt-Universität zu Berlin

Contact Information

Michael Seadle
Director and Professor
Berlin School of Library and Information Science
Humboldt University of Berlin
Unter-den-Linden 6
10099 Berlin
Phone: +49 (030) 2093-4248
E-mail: Seadle@ibi.hu-berlin.de

Websites

http://www.ibi.hu-berlin.de/forschung/digibib/forschung/projekte/LuKII

http://kopal.langzeitarchivierung.de/

THE METAARCHIVE COOPERATIVE

Profile

The MetaArchive Cooperative is a community-based network that coordinates low-cost, high-impact distributed digital preservation services among cultural memory organizations, including libraries, research centers, and museums. It was founded in 2004 with funding from the National Digital Information Infrastructure and Preservation Program (NDIIPP), and has also received funding from the National Historical Publications and Records Commission (NHPRC). Unlike vendor-based digital preservation solutions, the Cooperative enables cultural memory organizations to own and control the process of digital preservation for themselves.

The mission of the MetaArchive Cooperative is to foster better understanding of distributed digital preservation methods and to create enduring and stable, geographically dispersed "dark archives" of digital materials that can, when necessary, be drawn upon to restore collections at the contributing organizations. To these ends, the Cooperative runs a distributed digital preservation network and also consults regularly with groups that wish to form and host their own preservation networks.

The Cooperative is actively growing and welcoming new members. It is governed by a member-based Steering Committee and is hosted by the Educopia Institute, a not for profit organization that advances the production, dissemination, and preservation of digital scholarship and scholarly resources through fostering collaborative activities between libraries, museums, and other cultural memory organizations.

The MetaArchive Cooperative hosts subject- and genre-based preservation archives, including the Southern Digital Culture Archive, the Electronic Theses and Dissertations Archive (jointly hosted by MetaArchive and the Networked Digital Library of Theses and Dissertations), and the Early and Modern Literature Archive. Examples of collections that are being actively preserved in the preservation network are: Virginia Tech's born-digital works (e.g., electronic theses and dissertations, television news scripts, and the Faculty Archives) and scanned works from Special Collections, including university history, Civil War letters, and culinary manuscripts; Folger Shakespeare Library's digitized and born digital collections; and University of Louisville's uncompressed audio and image files from its digitized materials relating to Southern history and culture, including oral histories of African Americans and historic images of people, places, and crafts from Kentucky.

Participating Institutions

Auburn University, Boston College, Clemson University, Emory University, Florida State University, Folger Shakespeare Library, Georgia Tech, Library of Congress, Pontifícia Universidade Católica do Rio de Janeiro, Rice University, University of Hull, University of Louisville, University of North Texas, University of South Carolina, Virginia Tech

Contact Information

Katherine Skinner
MetaArchive Program Manager
Executive Director, Educopia Institute
Phone: (404) 783-2534
E-mail: katherine.skinner@metaarchive.org

Website

http://www.metaarchive.org/

THE PERSISTENT DIGITAL ARCHIVES AND LIBRARY SYSTEM (PEDALS)

Profile

The Persistent Digital Archives and Library System (PeDALS) is a project funded by the Library of Congress, National Digital Information Infrastructure and Preservation Program (NDIIPP) as part of its Preserving State Government Information initiative. This initiative focuses on capturing, preserving, and providing access to a rich variety of state and local government digital information. PeDALS uses LOCKSS software, developed by Stanford University Libraries, to maintain copies of the documents in separate physical locations and to provide automatic integrity and error checking.

In addition to developing a robust, trustworthy, inexpensive storage network, PeDALS aims to reengineer curatorial rationales to support an automated, integrated workflow to process collections of digital publications and records, as well as develop a professional network to promote collaboration and shared practices.

Participating Institutions

Arizona State Library Archives and Public Records, Alabama Department of Archives and History, State Library and Archives of Florida, New Mexico State Records Center and Archives, New York State Archives, New York State Library, South Carolina Department of Archives and History, South Carolina State Library and Wisconsin Historical Society.

Contact Information

Richard Pearce-Moses
PeDALS Principal Investigator
Arizona State Library, Archives and Public Records
Phone: (602) 926-4035
Email: rpm@lib.az.us

Website

http://www.pedalspreservation.org/

THE U.S. GOVERNMENT DOCUMENTS PLN (USDOCS)

Profile

The Government Publishing Office (GPO) is the official publisher of the U.S. Government and manages the Federal Depository Library Program (FDLP). They publish and distribute to libraries publications from 21 federal agencies as well as such integral publications as the Federal Register, Congressional Record, Congressional Reports, Bills, documents and Hearings, Public Laws, Papers of U.S. Presidents and much more. GPO Access—the online system that handles public access to these publications—is built on an older technology called WAIS with a very primitive user interface and limited search capabilities.

For that reason, the well known Internet- and open government activist Carl Malamud, with the assistance and cooperation of the GPO, harvested GPO Access documents from GPO servers in late 2007 and made them accessible/downloadable via BitTorrent, Rsync, HTTP and FTP. Those documents comprise 200+ gigabytes of data from 1991-2007 amounting to 5,177,003 PDF pages, 54,600 GAO Reports, 448,496 Congressional Reports and more.

Fifteen libraries in the U.S. Government Documents PLN (aka LOCKSS-USDOCS) have harvested the critical content from public.resource.org and are preserving it in a distributed digital preservation system. The group is currently planning to preserve documents from other sources including GPO's beta Federal Digital System (FDsys).

LOCKSS-USDOCS effectively replicates key aspects of the United States Federal Depository System (FDLP). The content is held in geographically distributed sites and replicated many times. Citizens thus have oversight and responsibility for the long-term care and maintenance of the content.

For more on LOCKSS-USDOCS, please see: http://freegovinfo.info/system/files/ATG-lockss-p5-7.pdf

Participating Institutions

Alaska State Library, Amherst College, Georgia Institute of Technology, Library of Congress, Michigan State University, North Carolina State University, Northeastern University, Rice University, Stanford University, University of Alabama,

University of Illinois/Chicago, University of Kentucky, University of Wisconsin-Madison, Vanderbilt University, Virginia Tech

Contact Information

James Jacobs
Government Information Librarian
Stanford Libraries
Stanford University
Stanford, CA 94305
Phone: (650) 725-1030
E-mail: jrjacobs@stanford.edu

Website

http://lockss.org/lockss/Government_Documents_PLN

Glossary of Terms

Administrative Server — Web host for the centrally managed resources of a PLN, such as the Title Database, Keystore, etc.

Archival Unit (AU) — An independent collection of content in a LOCKSS cache. Archival units are maintained as a whole by LOCKSS daemons. They are defined by the plugin and plugin parameters.

Cache — Server running LOCKSS software that stores harvested content on its local disk(s). Sometimes referred to as a LOCKSS Cache or LOCKSS Box.

Cache Manager — Web-based tool that helps monitor content on a network of LOCKSS caches, co-developed by the LOCKSS Team and the MetaArchive Cooperative.

Conspectus Database — A database tool developed by the MetaArchive Cooperative that holds collection-level metadata about the preserved collections.

Crawling — In the PLN context, LOCKSS incorporates a web crawler similar to those used by search engines, which follows chains of web links in order to retrieve copies of data from content contributors.

Crawl Rules — Defines the boundaries of an archival unit so that the LOCKSS crawler harvests everything the content contributor intends but does not harvest irrelevant content.

Daemon	Background Linux process which runs continually and executes software processes.
Dark Archive	Digital archive for which access to content is limited to organizational custodians.
Data Wrangling	Term used to describe the process of organizing archival units on a web server or a staging server so that they can be successfully ingested by the LOCKSS daemon's web crawls.
DDP	Stands for Distributed Digital Preservation. The process of creating copies of digital files and storing them in geographically distributed location for preservation purposes.
Dim Archive	Digital archive that incorporates elements of both the Dark and Open Archive models. Access for some materials is restricted to organizational custodians, while access for others may be open to a broad user community.
ETD	Stands for Electronic Theses and Dissertations, or digital files of masters theses and doctoral dissertations.
Firewall	Security element of a computer network that blocks unauthorized access and allows authorized communications.
Format Agnostic	Refers to a preservation software program that can accept any file format.
Format Migration	The process of converting a file from one format to another format, usually undertaken when a format is in danger of becoming obsolete.

Glossary of Terms

Harvest	Term used, sometimes interchangeably with "ingest," to describe the initial process undertaken by the LOCKSS daemon on a single Cache to obtain Archival Units from a content contributor's web server.
Hash	Also referred to as a checksum, this is a cryptographic signature for a file that enables comparisons of files. MD45 and SHA-1 are common forms of hashing. LOCKSS uses SHA-1 to help verify file integrity.
HTTP	Stands for Hypertext Transfer Protocol. A protocol that defines how files are transferred between servers and browsers on the world wide web.
Ingest	The act of crawling an AU and pulling the contents into a Cache.
IP Address	Stands for Internet Protocol. A numerical identifier assigned to every computer connected to the Internet.
JAR	Short for Java™ ARchive. A file format based on the popular ZIP file used to distribute a Java program.
Keystore	A file that is used to digitally sign a JAR file or to verify a JAR is signed by an authorized signature.
Kickstart	A method for automating Linux installations based on preconfigured settings.
LCAP	Stands for Library Cache Auditing Protocol. A very slow-acting network communication protocol developed for LOCKSS that allows the node caches to challenge each other in a polling process to prove that their copies of data are undamaged.

Linux	An open-source operating system that was designed to operate like the UNIX operating system.
LOCKSS	Stands for Lots of Copies Keep Stuff Safe. A software program (and public network) created by the LOCKSS team at Stanford University Libraries, originally for the preservation of electronic journals.
LOCKSS Daemon	Background Linux process which runs continually and executes the LOCKSS software processes.
MD5	A widely used cryptographic hash function. The MD5 value of a file can be used as a fingerprint to verify its integrity.
Manifest Page	Provides permission statement and a starting point for LOCKSS to crawl and harvest an Archival Unit from a data provider (publisher). The manifest page resides on the content web site.
Metadata	Information about information. In this context, typically this is descriptive information about a digital collection or digital file.
NDIIPP	Stands for National Digital Information and Infrastructure Preservation Program. A U.S. based program chartered by the Library of Congress to develop a national strategy to collect, preserve and make available significant digital content, especially information that is created in digital form only, for current and future generations.
Node	An individual LOCKSS Cache participating in a larger LOCKSS network.

Glossary of Terms

OAI-PMH	Stands for Open Archives Initiative Protocol for Metadata Harvesting, a protocol for harvesting metadata about digital collections residing in different repositories.
OAIS	Stands for Open Archival Information System. This is an ISO standard (or reference model) developed by the Consultative Committee for Space Data Systems for the acquisition, preservation, and dissemination of digital content.
Open Archive	A digital archive that is publically accessible.
Open Source	Software for which the source code is made freely available, usually under an open source license.
PLN	Acronym for Private LOCKSS Network. A LOCKSS network that is deployed by a set of like-minded institutions in order to preserve content in a closed preservation network.
Plugin	An XML file that instructs the LOCKSS software how to ingest and preserve content.
Plugin Repository	A storage space for jarred and signed plugins for collections in a LOCKSS network that is web accessible to each of the LOCKSS caches on the network.
Plugin Tool	A Java application developed by the LOCKSS team that simplifies the creation of new LOCKSS plugins.
Poll	A process performed by the caches in a LOCKSS network that involves the nodes communicating with one another regarding the integrity of the bits they are preserving.

Public LOCKSS	The public LOCKSS network preserves material of general interest to a broad community, including subscription-only material (largely e-journals), and is maintained by the Stanford University-based LOCKSS staff with funding provided by the LOCKSS Alliance.
RPM	Short for Red Hat Package Manager, a system for installing and managing software packages that is the standard for versions of Linux based upon RedHat Linux
SSL	Stands for Secure Sockets Layer. A method for encrypting data sent over a network.
SELinux	Security-Enhanced Linux is a set of mandatory access controls that can be applied to Linux systems to greatly enhance security.
SSH	Stands for Secure Shell, a command line tool to provide remote access and control to systems.
Subversion (SVN)	Utility for maintaining current and previous versions of source code files, plugins, scripts, and documentation.
Title Database	Central XML parameter file used to configure LOCKSS daemons.

Author Biographies

DWAYNE K. BUTTLER

Dwayne K. Buttler serves as the Evelyn J. Schneider Endowed Chair for Scholarly Communication at the University of Louisville in Louisville, Kentucky and holds a faculty appointment as a Professor in University Libraries. Much of his work focuses on the complex interrelationship of copyright law, licensing, and activities at the core of the university and library missions of teaching, learning, and scholarly communication. He earned a Doctor of Jurisprudence degree from the Indiana University School of Law-Indianapolis, teaches mass communication law at the University of Louisville, and leads frequent invited presentations on copyright and scholarly communication for administrators, faculty, librarians, students, and scholars in the library and the higher education communities.

MARTIN HALBERT

Dr. Martin Halbert is Dean of Libraries and associate professor at the University of North Texas. Prior to his current appointment at UNT, he worked in library administration positions at both Emory University (as Director for Library Systems) and Rice University (as a reference librarian and bibliographer). Halbert was an ALA/USIA Library Fellow stationed in Estonia assisting with the automation of the Tartu University Library. He is also president of the NDIIPP-funded MetaArchive Cooperative, an international consortium of research libraries and institutes that preserve digital archives in partnership with the Library of Congress.

RACHEL HOWARD

Rachel Howard is Digital Initiatives Librarian at the University of Louisville (Louisville, Kentucky). She holds an MLIS from the University of Washington and a B.A. in History from the University of Notre Dame. Prior to joining the University of Louisville Libraries faculty in 2006, she worked on the King County Snapshots, Smithsonian Global Sound, Global Performing Arts Consortium, and Library of Congress American Memory digital projects.

GAIL McMILLAN

Gail McMillan, Director of the Digital Library and Archives and Professor at Virginia Tech's University Libraries, joined the faculty in 1982 after working at the Smithsonian Archives and receiving Master's degrees in library science and history from the University of Maryland. Since 1994 she has led the Digital Library and Archives, also leading Special Collections for a decade, 1997-2007. Virginia Tech set the national and international standard for electronic theses and dissertations (ETD) and McMillan played a significant role in this initiative beginning in 1995. Since 2004 she has also represented Virginia Tech on the Steering Committee of the MetaArchive Cooperative. Following the tragic events of April 16, 2007, at Virginia Tech, McMillan began managing the online database of the scanned materials received from the worldwide community expressing their condolences. Throughout her career she has served in numerous national and statewide capacities and has presented and published about various aspects of digital libraries. Her recent publications and presentations focus largely on ETDs and in 2007 she received the Networked Digital Library of Theses and Dissertations' ETD Leadership Award.

MONIKA MEVENKAMP

Monika Mevenkamp has served as software engineer in disparate projects ranging from parallel computing to web development since she earned a M.S. in Computer Science in Germany. She immigrated to the US in 1989 where she worked at BellCore, Rice University, Georgia Tech, and Emory University. She currently leads technical development for the MetaArchive Cooperative.

BETH NICOL

Beth Nicol is an Information Technology Specialist with the Auburn University Libraries. Beth has worked with the startup of both the MetaArchive Cooperative and the Alabama Digital Preservation Network. In addition, during her more than 25-year tenure at the Auburn University Libraries, she has installed and managed multiple Library Management Systems and Digital Collections applications. She is a strong believer in Open Source software and sweat equity as well as the need for usable backups and the preservation of our digital (and digitized) heritage.

SUSAN WELLS PARHAM

Susan Wells Parham is Research Data Projects Librarian at the Georgia Tech Library and Information Center where she coordinates a cross-functional team responsible for auditing the research data environment on campus. She has over ten years experience implementing and managing digital library projects. Past projects include the Georgia Tech institutional repository, SMARTech, and the IPST Digital Collections. Susan's areas of expertise include metadata application and workflow design, information architecture, and technical project management. She has a B.A. in English Literature from the University of Virginia and an M.S. in Library and Information Science from the University of Illinois. Awarded the IMLS Digital Libraries Education Fellowship from the University of Illinois, she plans to complete the Certificate of Advanced Studies in Digital Libraries from UIUC in 2010.

BILL ROBBINS

Bill Robbins is the System Administrator for the MetaArchive Cooperative. He holds both a Master of Science and a Bachelors Degree in Electrical Engineering from the Georgia Institute of Technology. Prior to joining the MetaArchive Cooperative staff in 2008, he worked in Information Technology and Network Management at BellSouth, and as a design engineer for telecommunications companies in Florida and Georgia.

MATT SCHULTZ

Matt Schultz is the Collaborative Services Librarian for the Educopia Institute. Matt graduated Spring 2009 with a Master of Science in Information degree at the University of Michigan's School of Information. He specialized in Archives & Records Management, Digital Preservation and Human-Computer Interaction. During his graduate program Matt gained practical experience with digital libraries and archives through performing usability assessments on access systems, and designing scholarly web portals with open source software for the University of

Michigan Libraries and School of Information. His post-graduate work has involved performing trusted digital repository audits for LOCKSS and the MetaArchive Cooperative.

KATHERINE SKINNER

Dr. Katherine Skinner is the Executive Director of the Educopia Institute, a not-for-profit educational organization she helped to found in 2006 to act as a catalyst for collaborative approaches to the production and preservation of scholarship. She also serves as the Program Manager for the MetaArchive Cooperative, a distributed digital preservation solution for cultural memory organizations. She was previously the digital projects librarian at Emory University, where she served as co-PI on numerous projects in digital scholarship, access, and preservation arenas. She is one of the founders and the former Managing Editor of the *Southern Spaces* Internet journal and scholarly forum. Skinner received her Ph.D. from Emory University, and is the author of several articles, including "'Born Again:' Resurrecting the Anthology of American Folk Music" (*Popular Music*) and "The MetaArchive Cooperative: A Collaborative Approach to Distributed Digital Preservation" (*Library Trends*) and co-editor of two books, *Strategies for Sustaining Digital Libraries* (Emory University: 2008) and *The Guide to Distributed Digital Preservation* (Educopia Institute: 2010).

TYLER O. WALTERS

Tyler Walters is the Associate Director, Technology and Resource Services, Georgia Institute of Technology Library and is currently an ARL Fellow in the Research Libraries Leadership Fellows program. He is a co-Principal Investigator with the MetaArchive Cooperative (www.metaarchive.org), a partnership of the Library of Congress, National Digital Information Infrastructure and Preservation Program. Tyler serves on the Open Repositories Steering Committee and hosted the 4th International Conference on Open Repositories (https://or09.library.gatech.edu). He is also an interim governing board member of the Unified Digital Format Registry (UDFR) (www.udfr.org) and serves on Lyrasis' newly formed Digital Services Advisory Board (http://www.lyrasis.org/). Tyler teaches digitization and digital preservation at San Jose State University, School of Library and Information Science (http://slisweb.sjsu.edu/), created the curriculum for and teaches "Managing in the Digital Information Environment" in the Digital Information Management program, University of Arizona, and serves on the DigIn Advisory Board (http://digin.arizona.edu). Tyler served on the ARL/NSF workgroup that produced "To Stand the Test of Time: Long-term Stewardship of Data Sets in Science and Engineering" (http://arl.org/pp/access/nsfworkshop.shtml). He is on the faculty of the NEDCC's "Stewardship of Digital Assets" Workshop series and was a digital preservation consultant to the Library of Congress during 2009. Tyler is a frequent speaker, author, and recipient of the Society of American Archivists' Ernst Posner Award for best article in the *American Archivist* (1998).